Death Rays
and Delusions

Death Rays and Delusions

Gerold Yonas, Ph.D.

first chief scientist of Ronald Reagan's
Strategic Defense Initiative (SDI)
also known as Star Wars

with Jill Gibson

Peter Publishing

Cover design by Jenna Gibson
Cover art by Kinuko Craft

© 2017 Gerold Yonas, Ph.D. with Jill Gibson
All rights reserved.

ISBN: 0692919554
ISBN 13: 9780692919552
Library of Congress Control Number: 2017948347
Peter Publishing, Albuquerque, NM

Chronology of Events:

- *1960s: Basov and Prokhorov share Nobel Prize, Terra 3 laser program begins, reach megajoule*
- *1970s: Yonas starts fusion program at Sandia and in 1978 claims fusion in only 20 years*
- *January 1981: Inauguration of Ronald Reagan*
- *March 1981: Reagan assassination attempt*
- *November 1982: Brezhnev dies and Andropov accession to general secretary of Soviet Union*
- *March 23, 1983: Reagan Star Wars Speech*
- *June 1983: Fletcher Study, Defense Tech Study Team (DTST), Yonas leads DEW panel*
- *September 1983: Study delivered to Pentagon for review*
- *December 1983: Yonas asked by Gen. Bob Rankine to help prepare budget ($1.7 billion in 1985 and $3.7 billion in 1986)*
- *February 1984: Andropov dies and Chernenko accession to general secretary of Soviet Union*

- *April 1984: Yonas meets with Abrahamson at Challenger shuttle launch*
- *June 1984: Yonas accepts offer as chief scientist and acting deputy director*
- *June 1984: Army succeeds with HOE midcourse intercept after 3 failures*
- *December 1984: Gorbachev meets with Thatcher at Checquers to stop SDI*
- *March 1985: Chernenko dies and Gorbachev accession to General Secretary of Soviet Union*
- *March 1985: Yonas accused of aid to aliens at Pentagon meeting with Sen. DiConcini*
- *March 1985: Weinberger invites allies to participate in SDI*
- *May 1985: Yonas meets with Bush to prepare for SDI trip to Europe*
- *June 1985: TWA 847 highjack and Yonas trip canceled*
- *November 1985: Gorbachev/Reagan summit at Geneva*
- *December 1985: Abrahamson's wife dies in private plane crash and Yonas in charge of SDI for three weeks*
- *December 1985: Yonas represents Abrahamson; mumbles at Foreign Affairs Committee instead of lying*
- *January 1986: Challenger disaster viewed in Abe's office*
- *April 1986: Chernobyl disaster and Velikhov in charge of response*
- *August 1986: Senate votes to fund SDI at $2.7 billion instead of the $3.7 billion request*
- *August 1986: Adm. Nakhimov cruise ship disaster*

- *October 1986: Reykjavik summit ends, close to agreement to eliminate all nuclear weapons*
- *May 12, 1987: Gorbachev arrives at Baikonor to stop Polyus launch, but changes mind*
- *May 17, 1987: Polyus launched*
- *May 28, 1987: German private pilot flying from Finland avoids Soviet air defense and lands at Kremlin*
- *June 1987: Reagan in Berlin says "Mr. Gorbachev, tear down this wall."*
- *December 1987: INF treaty signed*

Contents

Chapter 1   A quest for a concept · 1
Chapter 2   Straight from science fiction · · · · · · · · · · · · · 8
Chapter 3   Beginning my path to SDI · · · · · · · · · · · · · · 13
Chapter 4   The dream of fusion · · · · · · · · · · · · · · · · · · · 22
Chapter 5   My friend the enemy · · · · · · · · · · · · · · · · · · · 29
Chapter 6   The Soviet mindset · 36
Chapter 7   Soviet laser technology · · · · · · · · · · · · · · · · · 44
Chapter 8   The view from the U.S. · · · · · · · · · · · · · · · · · 48
Chapter 9   Rudakov's real Soviet breakthrough · · · · · · · · 51
Chapter 10 The Star Wars speech · · · · · · · · · · · · · · · · · · · 55
Chapter 11 The Soviet reaction to the Star Wars speech · · · 59
Chapter 12 The Fletcher Study · 62
Chapter 13 The Scowcroft Commission · · · · · · · · · · · · · 76
Chapter 14 The SDI program takes off · · · · · · · · · · · · · · 82
Chapter 15 My SDI initiation · 87
Chapter 16 Life at the Pentagon · · · · · · · · · · · · · · · · · · · 94
Chapter 17 Humor can be risky · 97
Chapter 18 Dealing with the defense community · · · · · 107

Chapter 19 Facing our adversaries · · · · · · · · · · · · · · · · · 116
Chapter 20 An alien accusation · 122
Chapter 21 Edward Teller and developments in
           space wars technology · · · · · · · · · · · · · · · · · · 130
Chapter 22 Technical troubles · 136
Chapter 23 Budget battles · 143
Chapter 24 Adventures in public relations · · · · · · · · · · · 150
Chapter 25 Smart rocks and brilliant pebbles · · · · · · · · · 157
Chapter 26 Back in the USSR · 165
Chapter 27 Reaching out to our allies and the VP · · · · · 172
Chapter 28 The road to Reykjavik · · · · · · · · · · · · · · · · · · 183
Chapter 29 Setting the record straight in Tutzing · · · · · 190
Chapter 30 The Polyus launch · 196
Chapter 31 The beginning of the end · · · · · · · · · · · · · · · 201
Chapter 32 After SDI · 206
Chapter 33 A lesson and hope for the future · · · · · · · · · 212
Chapter 34 What next? · 217

Acknowledgements · 221
About the authors · 223
Endnotes · 227

CHAPTER 1

A quest for a concept

I WALKED INTO A SMALL windowless conference room buried in the depths of the Pentagon. On March 10, 1985, I had been called to a meeting in my role as chief scientist and acting deputy director of the Strategic Defense Initiative (SDI), popularly known as President Reagan's Star Wars program to defeat the evil empire, the Soviet Union, and eliminate the threat from nuclear weapons. My 20 years of experience as a scientist and engineer could never have prepared me for what was about to happen as I faced a room full of military colonels, a U.S. senator, a reporter and a well-known newspaper editor. After formal introductions and a confusing set of questions, the reporter turned to me and said in an accusatory tone, "Dr. Yonas, we have reason to believe you have been concealing an alien spacecraft on a Pacific island managed by the SDI. What do you have to say about that?"

In the many years that have passed since this strange encounter in the Pentagon, I have often reflected upon my involvement in the creation and evolution of the Star Wars beam weapon program and the part it played in the Soviet Union's demise. As chief scientist of the SDI, I participated in a complex quest to develop science fiction-esque space weapons and to decipher the political and economic factors that shaped decisions in the Soviet Union and the U.S.

From death rays to deception and disillusionment, I played a role in the scientific developments, political posturing and psychological battles that led to the end of the Cold War. These experiences gave me a unique, firsthand perspective on the scientific achievements and failures, the people, and the politics that shaped this time period. Today, I believe the technological developments that started with Star Wars are finally poised to become an integral part of minimizing the dangers and consequences of war, but it has taken decades to get to this point.

In the early '80s, America and the Soviet Union stood on the brink of nuclear war. After 20 years of the so-called "arms race," the Soviets had built up their nuclear weapons stockpile from 5,000 to 40,000, while at the same time, in the interest of detente, America had reduced its stockpile from 30,000 to 20,000. Many scientific and political leaders embraced the doctrine of Mutual Assured Destruction, or MAD, arguing that as long as both the superpowers had stockpiled sufficient weapons to obliterate one another in retaliation, even after a massive first strike, neither side would initiate a first strike due to the fear of devastating retaliation.

The concept of assured destruction as a method of preventing war originated in the '60s with the beginning of the

Cold War, when John Kennedy's secretary of defense, Robert McNamara, stated, "It is important to understand that assured destruction is the very essence of the whole deterrence concept. We must possess an actual assured-destruction capability, and that capability also must be credible. The point is that a potential aggressor must believe that our assured-destruction capability is in fact actual, and that our will to use it in retaliation to an attack is in fact unwavering. The conclusion, then, is clear: if the United States is to deter a nuclear attack on itself or its allies, it must possess an actual and a credible assured-destruction capability."[1]

While some leaders supported the theory behind MAD and the weapons build up that accompanied that belief, others sought a way to eliminate all nuclear weapons. In fact, John F. Kennedy remained thoroughly opposed to the weapons buildup, stating, "The weapons of war must be abolished before they abolish us."[2] Kennedy continued to push for nuclear test bans as a first step toward full disarmament. In his commencement address at American University June 10, 1963, he announced a new round of arms negotiations with the Soviets. "If we cannot end our differences," Kennedy said, "at least we can help make the world a safe place for diversity." He challenged Russia to a "peace race" instead of an "arms race" and once remarked, "It is insane that two men, sitting on opposite sides of the world, should be able to decide to bring an end to civilization."[3]

Many years later, in 1982, the well-known Princeton physicist Freeman Dyson delivered a series of lectures about the threat of nuclear war. Dyson summed up the situation by stating, "We now possess weapons of mass destruction whose capacity for killing and torturing people surpasses all our imaginings. The Soviet government has weapons that are as bad or worse. We have been almost totally unsuccessful in halting the multiplication and proliferation of these weapons." Dyson went on to advocate persuasively for a negotiated move from nuclear weapons-based retaliation toward increased reliance on defense-based deterrence. He suggested that a solution might lie in developing "a concept of weaponry which would allow us to protect our national interests without committing us to threaten the wholesale massacre of innocent people."[4]

Dyson was known as a creative iconoclast. He reminded me of the wise character Yoda in the *Star Wars* movies. I once met him at a professional conference during which he told me about a paper he had written in 1980 that called for a shift from an ever-increasing competition of offensive weapons to a competition of increasing defenses. Dyson saw the change as being one characterized by a cooperative transition, and not more competition. In the paper, titled "The Quest for a Concept," he wrote, "If we decide on moral grounds that we choose a defensive world as our long range objective the political and technological means for reaching the objective will sooner or later be found, whether the means are treaties and doctrines or radars and lasers."[5]

Dyson's proposed "live and let live" philosophy appealed to me, but the vision seemed unlikely as the United States and the Soviet Union hurtled ever closer to war. I would revisit Dyson's words many times over the next few years as I became embroiled in the political and technological battles that surrounded the conflict between the U.S. and the Soviet Union, the Strategic Defense Initiative and the issue of arms control.

Like Dyson, America's president, Ronald Reagan, aimed to eliminate the threat of nuclear war. In the early 1980s, Reagan was seeking to build up U.S. military capabilities to deal a fatal blow to what he deemed "Godless communism."[6] He called for a freeze in growth of all government spending except for the defense budget – that he wanted to expand drastically.

Reagan's major strategic focus was putting pressure on the Soviet Union by threatening the deployment of intermediate range missiles in Europe. These missiles could reach Moscow in minutes with little or no warning. Reagan's strategy worried the Soviet military and Politburo leaders and sparked a growing outcry from many decision-makers throughout the world. Critics feared this U.S. missile buildup would create dangerous instability between the two global superpowers.

In 1983, there was a feeling of growing economic optimism in the U.S. The American economy had started to improve after the recession of the early '80s. The price of oil was going down and the Dow was hitting new highs every month. The other thing that was on an upsurge was the growing anti-nuclear and anti-Reagan sentiment throughout the world. Reagan wanted to accelerate the arms race and, based on information from his close friend and head of the CIA, Bill Casey, he had confidence that the Soviet Union could not compete with our military buildup. Meanwhile, the American president was becoming less and less credible in elite intellectual circles. Reagan's detractors in academia thought he was a good actor, but a likable dunce.[7]

Yuri Andropov, the aging leader of the Soviet Union, was doing everything he could to put a stop to the Reagan defense buildup, but attempts at arms control were failing, and eventually the Soviets pulled out of the Geneva arms control talks. The years of effort to achieve arms control were coming to an end and, as a result, the next step had to be a renewed investment in strategic arms on both sides. The Soviets had deployed large ICBMs with multiple warheads that threatened the United States' retaliatory capabilities, and America's next step would be building more offensive weapons to counter that Soviet buildup.

The situation sparked fear throughout the world. Hundreds of thousands demonstrated throughout Europe against the proposed U.S. nuclear buildup. Andropov attempted to capitalize on this discontent and traveled to Europe to argue against the continuation of the nuclear arms race. Fear also swept through the U.S., where people were both concerned and confused by media discussions of "nuclear winter" and predictions of doom. In 1983, the made-for-TV movie *The Day After* portrayed a Soviet nuclear attack on a Midwestern U.S. city and the subsequent nuclear holocaust. People began lobbying for a nuclear freeze, and the anti-nuclear movement got a boost from the publication of a Catholic bishop's Pastoral Letter on War and Peace that expressed "profound skepticism about the moral acceptability of any use of nuclear weapons." The Soviets condemned Reagan for being unwilling to join them in their quest for a "safer world."[8]

Andropov, with a serious heart condition, was no match for Reagan and the improving U.S. economy.[9] Our two countries were clearly moving in different directions, away from the period of detente and toward more and more confrontation. The relationship between the two superpowers had deteriorated, arms control negotiations had hit a dead stop and America's leaders were ready to consider a new idea – a new type of weapon – and the concept was to come straight from the pages of a science fiction novel or movie script.

CHAPTER 2

Straight from science fiction

THIS QUEST FOR A FUTURISTIC weapon to achieve a military advantage was first described in science fiction. In *The War of the Worlds*, published in 1897, H.G. Wells wrote of the use of "death rays – An almost noiseless and blinding flash of light ... the unseen shaft of heat passed over them ... sweeping round swiftly and steadily, this flaming death, this invisible, inevitable sword of heat."[10] Wells' idea of a speed of light energy weapon captivated the science fiction buffs like me in the '50s, but at that time there existed no way to actually build his fictional device. In 1925, Russian novelist Alexi Tolstoi revived this notion of a beam weapon in his book *The Garin Death Ray*.[11] Tolstoi captured the imagination of the Soviet military by describing "transmission that does not disperse ... to cut through a railway bridge in a few seconds." Tolstoi also characterized his laser-like weapon as "an invention that smells of higher politics" and noted that "our enemies must not get it."

Tolstoi's book was filled with speculation about beam weapons along with prophetic thoughts about the mind of the fictitious weapon designer, Garin. "He's ambitious. An absolutely isolated individual. An adventurer, a cynic with the makings of a genius. Too much temperament. A monstrous imagination. But that wonderful mind of his is always motivated by the lowest desires."

The weapon itself was described as transmitting "infrared rays over a distance … heat waves at a temperature of a thousand degrees centigrade transmitted parallel to each other." Garin explained, "I can reduce the ray cord to the thickness of an ordinary needle." When I read this, I thought of many of the present day publicity quotes about high-power lasers, but when this was written, the invention of the laser was still decades away.

In 1929, the comic strip *Buck Rogers in the 25th century A.D.* debuted featuring a science fiction ray gun. The toy version of the gun appeared in 1934 and retailed for 50 cents.[12] From *Lost in Space* to *Star Trek*, lasers, phasors, blasters and other directed-energy weapons have played a major role in comic books, movies and television, but the death ray was simply science fiction – at least at first.

Personally, I was captivated by science fiction and became a dedicated fan at a young age. I remember attending the movies with my older brother. We would walk to the Shaker Theatre Saturday afternoons. With a quarter from my mother, we could buy two 10-cent tickets and still have five cents left over for a box of Jujubes candy that glued our teeth together – making them nearly impossible to chew. I specifically recall seeing *The Day the Earth Stood Still* at age 13. The movie featured an alien who, accompanied by a death-ray shooting robot, comes to earth to save the planet from war. Naturally, the film ended with dire warnings of death and destruction while prominently featuring beam weapons that wipe out everything in their path. Re-watching the film as an adult, it struck me that a powerful theme of the movie is the communication breakdown between the humans and the alien. Produced at the height of the Cold War, the science fiction classic also provides a powerful anti-war message. Director Robert Wise remarked on this, saying, "I feel very strongly in favor of what the movie says. It's very much a forerunner in its warning about atomic warfare, and it shows that we must all learn to get along together."[13] At 13, I wasn't thinking about the political implications of the movie, but now I can't help but wonder if somehow *The Day the Earth Stood Still* influenced my approach to strategic defense.

As a movie-going boy, I did not realize that the development of "real" death ray technology was already underway. Scientist Nikola Tesla gets my credit for developing a theory behind the real death ray. Tesla, the inventor of AC power, was an obsessive, frustrated, wildly creative loner. In many ways, he reminded me of the fictitious Garin. Tesla scoffed at the inventions of infrared or radio waves, which he said could not be a true beam, but instead proposed a particle beam, "of microscopic dimensions, enabling us to convey to a small area at a great distance trillions of times more energy than is possible with rays of any kind … nothing can resist."[14] He claimed his invention could "destroy anything, men or machines approaching within a radius of 200 miles." Tesla's claims echoed his era's familiar science fiction, but he was the first to develop a concrete scientific proposal.

In the 1930s, Tesla began seeking money to build his conceptual death ray. He asked J.P. Morgan for funding, claiming, "The new means I have perfected affords absolute protection against aerial bombing," but the money was not forthcoming. He then tried Neville Chamberlain of Great Britain and the League of Nations, but had no luck. Eventually he turned to the Soviet Union. In 1937, Tesla found his first paying customer when he sold the particle beam idea to the USSR for $25,000. By then the 81-year-old inventor on the edge of poverty, with no friends other than his pet pigeons, was at the end of his rope, but the Soviet Union had the glimmer of the first particle beam weapon concept. Tesla, a true pacifist, had invented the "weapon to end all wars."

With the arrival of World War II, the technical community shifted its attention away from beam weapons and focused on the development and deployment of nuclear weapons, ballistic missiles and ballistic missile interceptors. Particle beam weapons dropped from view but resurfaced when the gleam in the eye of science fiction buffs, military dreamers and some scientists and engineers became a reality with the invention of the laser in 1961. The key property of a laser beam was its very low divergence, opening up the realistic possibility of the true death ray weapon as envisioned by Garin in the 1920s and Tesla in the 1930s.

Reagan himself gained firsthand exposure to the concept of death rays long before he began his political career. In fact, he was still working as an actor. One of the most prophetic roles of Reagan's movie career was his depiction in 1939 of Secret Service agent Lieutenant Brass Bancroft in a series of action-adventure B movies. In one of those films, Reagan, as Bancroft, saved the "death ray projector" that could make the U.S. invincible in war and became "the greatest force for world peace ever discovered."[15] Perhaps this experience influenced his strategic policy, as he became the leader of the free world.

So both science and science fiction had paved the way for the creation of an energy weapon that could serve as the ultimate super-weapon, capable of annihilating anything in its path. In the 1980s, the SDI would turn the death ray from an offensive weapon to a defensive weapon. As author Benjamin Wakefield noted, "Seemingly lifted straight from the pages of a comic book, the program suggested that a network of satellite-based lasers could be used to destroy missiles in flight."[16] I find it both amusing and ironic that some of the most important work I would pursue in my scientific career would bear an uncanny resemblance to the weapons and plotlines of "B" movies and comic books.

CHAPTER 3

Beginning my path to SDI

My high school biology teacher, Mr. Linshield, was a small, bald man with horn-rimmed glasses and an apologetic manner. He scurried from class to class with his head down and rarely raised his voice. That is, until one day when he fell victim to one of my first experiments. I was in the 11th grade, and the success of that experiment sparked (pun intended) my entire career.

In preparation for the upcoming science fair, I had built a Van de Graaff generator – a simple electrostatic machine that produces a very high static voltage. I had developed a habit of leaving the generator charged up in the chemistry classroom so I could enjoy the shrieks of the unsuspecting girls who walked too close to the machine, causing their hair to stand on end. One day, however, it was Mr. Linshield who walked too close to the generator, causing a dramatic blue and orange lightning bolt to jump from the machine to his bald head.

"Yonas," bellowed the teacher, reeling from the static discharge. Naturally, I was nowhere to be found. Later that year,

I would win second prize in an Ohio state science fair for an experiment I devised using the generator. It was my first chance to create, conduct and explain an experiment, and I immediately knew this was what I wanted to do with the rest of my life.

In addition to winning a science fair trophy, which was actually just a water glass adorned with the state of Ohio seal, and mastering the science behind my project, I enjoyed explaining my experiment to the contest judges. I loved the attention that came with being an instant expert, but even more importantly, I embraced the discovery that I could make my audience laugh. My sole knowledge of comedy came from the few times during my boyhood when my parents scraped up enough money to take me and my four brothers to the Catskills resort Kutsher's Hotel and Country Club in the so-called Borscht Belt of New York. At Kutsher's we saw Jewish comedians, including my personal hero, Shecky Greene, and I dreamed of becoming a stand-up comic. Combining comedy and science provided the natural way for me to live two of my life dreams simultaneously, and I soon found that I would rely on my ability to find humor in difficult situations throughout my life.

My science fair success led to college, where I went on to earn degrees in engineering physics at Cornell University in 1962 and engineering science and physics at Caltech in 1966. My background focused more on engineering than on physics and, although my studies in college were rather eclectic, they were not oriented toward the fields of powerful beam weapons. Instead, much of my early career was focused on magnetohydrodynamics (MHD), the study of the properties of the motion of a conduction fluid in a magnetic field. I even won a prize for an undergraduate paper titled "An Induction Electromagnetic Pump," which would later play a role in the allegations that I had knowledge of alien spacecraft technology.

Yonas graduates from Caltech in 1966 with his Ph.D. in engineering science and physics.

While completing my Ph.D. at Caltech, I was fortunate to be able to do my experimental thesis research under the very inspirational professor Hans Liepmann. My experiment used a very special wind tunnel, developed by Tony Maxworthy at the Jet Propulsion Laboratory, in which sodium was heated to a liquid state. My most exciting memory from this research was the time the fire marshal walked into the lab while I was lying on my back tightening a leaking fitting that was dripping liquid sodium into a tray filled with kerosene. Naturally, this represented a major fire hazard, but I managed to talk my way out of being fined or – even worse – having my experiment shut down. The fire marshal asked a few questions, and I answered in technobabble – a technique of verbal obfuscation that I would later perfect and use frequently when working in the Pentagon. In any case, my jargon-filled, high-tech explanation managed to confuse – and defuse – the fire marshal. He left, and I continued my work. In today's world of safety rules, this incident probably would have been the end of my research. As it was, I finished my experiment, published the research, graduated with my Ph.D. and soon landed a full-time job at NASA's Jet Propulsion Laboratory (JPL), where the first lucky break of my career would soon stem from someone else's failure.

In the early 1960s, JPL was focused on conducting unmanned space missions to obtain images of the surface of the moon. While I was not involved with JPL's Ranger moon exploration program, it would shape my fledgling career. The Ranger spacecraft were designed to crash onto the surface of the moon and capture photographs, but, when I started at the lab, JPL had conducted several Ranger missions and all of them had failed. Because I was not working on this area, I wasn't concerned about the Ranger mission flop, but my colleagues joked that the reason Ranger could not successfully capture photos was because "Yonas forgot to remove the camera lens cap before the launch." After the sixth unsuccessful Ranger mission, the JPL administration decided to eliminate most of the employees who were not contributing to the Ranger program. Since I was working on other projects, I was told to find another job. This was the first major setback of my life but in the end, it turned out to be one of the greatest opportunities in my career.

Getting laid off from my first real job after spending nine years in higher education taught me some very important lessons and shaped my career goals. I decided that in the future, I would be very careful when choosing employment and would consider the organization, the role and the position within the company very carefully. I wanted to make sure to be in charge of my own fate; to be the boss rather than the employee; and I learned to always pay attention to project and organization plans and budgets. I also decided that five years spent working on the same area of research was more than enough. I vowed that, from then on, I would move on after five years devoted to a particular subject.

As I began my job search, I had opportunities to join large aerospace companies in L.A. to work on high power lasers, which were becoming more and more popular areas of development for futuristic weapons. Instead, I chose a small company, Physics International, or P.I., in the San Francisco Bay area recommended by a friend from college and grad school, Alan Klein, who worked there. At P.I., I would work on a new subject of which I had no knowledge or relevant experience. All I knew was that the project involved high voltage generators, and I had a feel for "arcs and sparks" from my high school science fair project. Young and optimistic, I believed I could make things work, and even more importantly, I would be on the ground floor in a small company. I believed I would be in charge of my own destiny.

At P.I., I would also be experimenting with the new and rapidly emerging field of high power, relativistic particle beams driven by pulsed high voltage generators. The U.S. Department of Defense was funding the study of high voltage pulsed power as a strategy to test how well our country's nuclear warheads could survive the Soviet ballistic missile defense. American defense technology had become reliant on the sophisticated electronic systems on nuclear warheads, and our scientists needed to determine whether the American electronic systems could survive the high-intensity radiation produced by the Soviet nuclear explosive-powered defense. The goal was to create an affordable way to test our systems without having to conduct expensive underground nuclear explosions. So the Defense Department set out to develop giant high power electron beam machines to produce intense radiation to test our electronics. Physics International, my new employer, had the contract to develop these machines.

While the original intent of the pulsed power technology was to use the pulse to simulate the bursts of radiation from exploding nuclear weapons, the research also held the promise of providing a way to generate the enormous amounts of energy needed to create nuclear fusion. Like many of my colleagues and other physicists before me, I was fascinated by the possibility of harnessing the process that powers the sun to create an unlimited source of clean, cheap energy. Fusion is a nuclear reaction that releases energy by fusing light nuclei, together providing a reliable source of energy without the dangers of fission-powered nuclear reactors. At P.I., I was charged with finding a way to test the missile re-entry vehicles, the part of the missile that re-enters from space on the way to its target. To do that, I would need to find a way to focus an intense relativistic electron beam onto a BB-sized spherical target. The concept was to implode the spherical pellet, thus compressing and heating a small amount of thermonuclear fuel. In effect, the goal was to create a small fusion explosion in the laboratory. But my first goal was to figure out how to control the compressed pulse of energy generated by the powerful electron beams from the pulsed power machines. I soon discovered that this was easier said than done.

When I arrived at P.I., the company had recently modified a pulsed power machine to produce very high currents. Unfortunately, the scientists had not been able to use this machine, called the Big Blue Boy, to generate a high power electron beam. Big Blue Boy was designed to deliver a high voltage pulse of one hundred kilovolts that lasted less than one hundred nanoseconds. The rapidly rising pulse would hit the vacuum gap between the negative electrode, the cathode and the positive electrode, the anode, causing the surface of the cathode to explode into a plasma that emitted electrons that were accelerated across the vacuum gap to the anode. The result would be the intense beam that would simulate the impact of radiation from a missile defense nuclear warhead. The problem was that I couldn't get it to work.

With a budget of $50,000, I began experimenting with the high power beams at levels that had never existed before. There was no room for error, and my salary depended on my ability to master this task. After a few months on the job, and after intensive theoretical study of the physics of high current relativistic beams, I had achieved absolutely no success – even working with a fairly low power electron beam from the smallest pulsed power machine we had. After losing my first job, I faced the possibility of failure at my second. Frustrated and depressed, I went down the hall one day to talk to the company dreamer, Dave Sloan, who had worked with Earnest O. Lawrence on the Manhattan Project. Sloan was in his 70s and kept a cot in his office for afternoon naps. Fortunately, he was not napping when I dropped by seeking his advice. Sloan showed me a sketch of something that looked like a giant disk-shaped object that he called "the whip." He described the flow of power as the electrical pulse imploded slowly from the outside of the disk and accelerated toward the central axis and then finally converged onto the BB-sized fusion fuel capsule. Sloan said it would be like a whip where you move the handle slowly and the tiny tip of the whip cracks the sound barrier in the surrounding air. He explained, "When you push on micrograms with mega amps, you can make fusion." Reenergized, I went back to my lab.

Back in the lab, the beam still was not cooperating. It would not even do the same thing more than once, and I was rapidly exhausting my meager budget. Instead of producing a high power beam, the experimental result was not repeatable and whatever beam that was produced was totally unstable. All I had managed to create was mostly a short circuit arc in the diode. I was about to short circuit my career due to the rapidly declining funds and the frustration I felt. Fortunately, I had a colleague, Don Pellinen, who was an electrical engineer and knew far more than I did about pulsed power. Pellinen felt the problem was caused by premature firing of the machine before the "whip was ready to crack." He went off and came back in an hour holding in his hand a simple but elegant piece of conical Plexiglas that he had turned on a lathe. The piece was called a "pre-pulse" switch designed to hold off the voltage until the beam was ready to go. The "pre-pulse switch" did the job, and the very next shot was perfect. I had gained a profound appreciation for how first class engineering was necessary to make physics a reality. Soon my colleagues and I had tamed the beam and we were able to begin the Department of Defense's radiation tests. My career as a beam physicist and fusion scientist was on its way – saved by a $1 piece of plastic.

CHAPTER 4

The dream of fusion

PLUNGED INTO THIS WORLD OF nuclear weapons and high power beams, I remained intrigued by the ability to use that same powerful electron beam technology to create nuclear fusion. My curiosity soon led to an obsession with the goal, and I began to seek a way to pursue the challenge of creating fusion with a much bigger budget. I was aware of a similar high power beam research program going on at Sandia National Labs in Albuquerque, New Mexico, which was funded by the Atomic Energy Commission (AEC) rather than the Department of Defense. It seemed to me that the research "grass" was growing a lot taller with AEC funding, and with my limited understanding of the changing national military strategy, I concluded that I could have access to additional funding if I moved to Sandia Labs. I also felt the growing importance of ballistic missile defense would lead to greater investments in large nuclear weapons effects simulators, which would mean new opportunities for high power beam research. Sandia was developing all the complex electronics in nuclear weapons, and those electronic systems all needed testing and certification using high power beam radiation, so I was certain the lab would have growing budgets to satisfy this need. Plus, I surmised, I could continue to pursue fusion research as a spinoff, so moving to Albuquerque was the natural next step.

But following my dream would mean transplanting my family. In 1961, I had married my high school sweetheart, Jane, and my first child, a daughter named Jill, had arrived in 1966 along with my Ph.D. Jane grudgingly supported my desire to forsake the San Francisco Bay Area for the deserts of New Mexico – perhaps because she had never visited Albuquerque. I truly appreciated her willingness to move to a city where she often said "the only green things were the freeway signs" and the best store in town was JC Penney. Nevertheless, despite the dry desert air and the lack of sophistication, Albuquerque would soon feel like home to us both.

I arrived at Sandia in 1972, prepared to spend large sums of money on weapons effects simulators, which would allow me to pursue my interest in fusion research, but I immediately hit a roadblock. The U.S. and the Soviet Union had just decided that ballistic missile defense was not such a good idea after all, and they had signed the 1972 ABM treaty. The two nations had committed to the Mutual Assured Destruction doctrine and had concluded that ballistic missile defense would destabilize that arrangement. That meant the funding I was counting on would no longer be available. The nuclear weapons budget at Sandia was dropping, and this budget reduction soon resulted in the lab's first layoff in its history, which took place shortly after I came on board. Luckily, I survived my second layoff experience, but once again I faced a change of plans due to something completely beyond my control. I needed to start scrambling. With funding for defense and weapons research declining rapidly, I had to find a new way to fund my dream of experimenting with fusion technology.

Fortunately, there was another application for my research – and it was trendy. The energy crisis of 1973 had increased the government's interest in alternative energy projects. The national laboratories were under pressure to identify secure, environmentally safe sources of energy that would reduce dependence on foreign resources, and harnessing fusion for energy promised a possible option. Fusion research was already underway at our rival national labs, Lawrence Livermore National Laboratory (LLNL) and Los Alamos National Lab (LANL), which had begun programs that relied on lasers to develop fusion and focused on weapons applications. Despite the LLNL and LANL head start, a small group of colleagues and I created a proposal for funding from the AEC fusion division, and we received money for a small study on electron beam fusion research. Later I learned that a rivalry between the weapons branch and the energy branch at the AEC had contributed strongly to the funding of our research. Our successful attainment of funds was based on politics – not the brilliance of our proposal or the strong scientific basis behind our program – but I didn't care. I had the money and support to pursue fusion research, and nothing could stand in my way.

I spent the 1970s and early '80s embroiled in fusion research, learning about particle beam energy and high current proton beams and pulsed power, which would ultimately shape my approach to strategic defense. But I was not the only scientist pursuing the dream of fusion. Similar experiments were going on at other U.S. laboratories and at hidden facilities in the Soviet Union. Sandia Lab's two rivals, LLNL and LANL, both were researching the use of lasers to generate fusion – and one of the leading scientists in this research was Edward Teller, the father of the hydrogen bomb. Little did I know that I would soon begin working closely with the man known by many as "the real Dr. Strangelove."

Teller, the brilliant and controversial Hungarian immigrant who worked on the Manhattan Project, had become a fixture at Lawrence Livermore, where he advocated strongly for developing nuclear fusion for both energy and weapons purposes. Unlike others at LLNL who focused on competition, Teller had maintained a collegial interest in Sandia's fusion research and, in the mid-1970s, he asked me to write a chapter for a book he was compiling on fusion research. After I submitted the chapter, Teller requested that I meet him in Los Alamos to discuss the research. What followed was one of the most stimulating technical conversations of my career. We spent a whole day together discussing physics equations and scribbling on a blackboard. Teller was very interested in our fusion experiments at Sandia and had numerous suggestions and comments. Like a professor, he led me through an invaluable process of explaining and organizing my thoughts. Shortly after our meeting, he announced plans to visit my lab.

Teller came to Albuquerque on a Saturday, and we started the tour with breakfast at my house. This meant a planned family excursion to the lake had to be delayed, much to the annoyance of my 3-year-old daughter. Teller was not offended by her complaining, saying in his thick Hungarian accent that he understood because he had children, too. While Teller seemed to understand my daughter's behavior, he appeared perplexed after viewing the electron beam fusion accelerator prototype we had built. Although he clearly understood the physics behind our experiments, he found the actual hardware and engineering confusing. "You engineers are very hard to understand," he said, shaking his head. I realized then the profound difference between theoretical physics and practical applications. Teller was a brilliant scientist, but he wasn't an engineer. He much preferred thinking about the concepts to actually making them work.

Incidentally, the book Teller asked me to contribute to would never be published. The Atomic Energy Commission decided the publication could threaten national security due to Teller's connection to the "secret of the H-bomb," also known as the Teller-Ulam principle. Although the prospective book did not actually discuss this classified concept behind creating a fusion-powered hydrogen reaction, which at that time was one of the nation's most heavily guarded nuclear secrets, the U.S. government wanted to avoid any possibility of revealing a secret that could give other countries the power to build their own thermonuclear weapons. Despite these precautions and unbeknownst to the U.S. government, the Soviet Union already had the information and was using it in their weapons program. Curiously, after years of legal wrangling, the basic principle behind the hydrogen bomb was later published in the popular press after it was revealed by a Soviet plasma physicist, Leonid Rudakov, but more about him later.

The Sandia program moved along with a strong focus on electron beams but with a backup plan of switching to ions if that appeared to be a more favorable approach. Although the focusing seemed to be more achievable with electrons, we were concerned about their ability to penetrate deeply into the target, which made target design rather iffy. We knew that we had several ways to generate high current proton beams, and eventually the target concepts for ions made that the desired approach. We received permission and funding to build the $14 million Electron Beam Fusion Accelerator (EBFA) and began testing. When we were convinced that we could focus the ions well enough, we changed the name of the machine to the Particle Beam Fusion Accelerator (PBFA) and requested $56 million for a much larger machine.

Meanwhile, laser fusion research continued at LLNL and LANL, where the emphasis remained on applying weapons concepts. In the late '70s, both labs were developing a new concept they dubbed the Third Generation Nuclear Weapon, which would use the output of a nuclear explosion to power an X-ray laser that would allow long distance precise attack on a target in space. Because of my area of research, Sandia management assigned me to investigate the subject and write a report on the weapons/fusion work going on at LLNL and LANL. Teller was a strong proponent of the X-ray laser concept, which he was marketing aggressively as the mechanism behind a space-based nuclear weapon-driven ballistic missile defense weapon. While I had already been involved in several studies on various ballistic missile defense approaches, I saw nothing particularly promising in his approach. Although the physics of the X-ray laser was very challenging and created substantial interest, the practical application of the concept seemed to me to be out of the question. Furthermore, when it came to making fusion energy a reality, I favored the more efficient and cost-effective particle beam approach, which seemed more practical and easier to engineer. I reported back to Sandia management that the laser-based concept under development at LANL and LLNL was exciting but impractical. Also, with the 1972 ABM treaty prohibiting future ballistic missile defense development and deployment, I saw no reason for Sandia to waste its time with weapons-oriented, laser-based research. I was in the business of creating an unlimited source of clean, cheap energy – not building weapons. At least that's what I thought.

CHAPTER 5

My friend the enemy

THE YEARS FROM 1972 TO 1982 kept me busy learning about particle beam fusion and its energy and weapons applications – but I was also studying something else. My work on fusion research at Sandia allowed me to interact with Soviet scientists who were pursuing similar research in their country. As I exchanged information and attended international conferences with my Soviet counterparts, I observed their behavior and culture. I really wanted to know what made them tick. I would later draw upon this knowledge of the Soviet Union when I worked in the Pentagon, and I believe my exposure to and understanding of Soviet culture strongly influenced my thoughts regarding Soviet/U.S. weapons negotiations and strategy. During my years at Sandia however, I never considered the Soviet citizens as representatives of an evil empire. Many Soviet citizens had become my close friends.

Like me and my co-workers, a group of Soviet physicists was conducting fusion research, and we often crossed paths at international conferences on the topic. Our government strongly encouraged us to engage with the Soviet scientists – seeing our interactions as a way to learn more about secret Soviet weapons programs. Despite this hidden agenda, I soon discovered that these physicists from the other side of the world shared my interests, my sense of humor and my intense drive to unlock the secrets of igniting fusion fuel in the lab. As a result of our mutual interests, I began a series of technical exchanges with a research group at the Kurchatov Institute in Moscow, headed by Leonid Rudakov, a 40-year-old theoretical plasma physicist. Rudakov's friends called the athletic and brilliant scientist Lonya. I called him my chief global competitor in the race to create fusion – and my friend.

The Kurchatov Institute served as the leading open nuclear lab in the Soviet Union. Two other closed labs were dedicated to nuclear weapons, but my Soviet colleagues never admitted their existence to me. Instead, I made many trips to Moscow and also visited facilities in Novosibirsk, Siberia and Leningrad. I spent most of my time with Rudakov, but I sometimes also interacted with his boss, Evgeny Velikov, who enjoyed egging on the competition between me and Rudakov to motivate his countryman to work harder. Not that Rudakov needed pushing – much like me, he was obsessed with fusion's promise of peaceful and plentiful power. In many ways, Rudakov and I functioned more like allies than competitors, as we both were on a quest to get support and financial backing for our respective research programs. We were interested in particle beams, and both our countries were giving greater funding to the laser approach. Our technical discussions benefited both of us and the competition kept us on our toes, but the issue of weapons applications always hovered in the background. Together we played a complex game of openly talking about fusion but avoiding directly addressing the related weapons applications of our research. At times, we baited each other with probing remarks and indirect questions. The game was to uncover the unknown without ever admitting that the game was on. I always tried to understand not only what the Soviet scientists said, but what they were thinking and doing. I quickly learned that the Soviets were good at keeping secrets as a way of life. Perhaps this was a habit learned during the Stalin years. The only times the Soviet scientists seemed to be forthcoming was when there was no conceivable way for anybody else to be listening, such as when we were walking in the woods. Sometimes a program leader would have a few glasses of vodka too many and would become more communicative, but when I began to probe, my friends would quickly change the subject or launch into a Russian ballad and lead the group in song. I quickly discovered that my Russian colleagues could hold their liquor. They told me, "Gerry, you drink like a child," because I avoided drinking much vodka. Many nearby potted plants were irrigated by the contents of my vodka glass when my colleagues looked away.

Despite the teasing, my Soviet colleagues were always exceptional hosts. I recall one visit especially fondly. We had traveled to the National School on Plasmas in the "science city," Akademgorodok, near Novosibirsk, Siberia. I, along with a few Americans, had been invited to this Soviet meeting arranged by Serge Budker, who created the Institute of Nuclear Physics, one of the leading nuclear research institutions in the Soviet Union. Budker was known for his brilliance, creativity and open mind to new ideas that often rubbed the bureaucracy the wrong way. The highlight of the trip included a private lunch with Budker, and I recall asking him about the work of a Soviet scientist named A.I. Pavlovsky, who was not known in the West but had published fascinating papers on my field of pulsed power and accelerators. I suspected Pavlovsky worked at a closed nuclear weapon laboratory, and Budker's response was, "Your spymasters told you to ask this question!" I sputtered something stupid, but I later found out that my suspicion had been correct.

Another memorable event was a banquet on a small island in the middle of the Ob River, which was also called the Ob Sea due to its enormous size. Our host, Budker, arranged to have the hundreds of meeting attendees transported in a large ship that could not dock at the uninhabited island. When we departed, Budker was not to be found, but on our way to the island, it was announced on the ship that we were being attacked by pirates. As we looked over the side we saw Budker's private speed boat, "The Proton" flying the Jolly Roger, and we stopped so the pirate could board and join us. When we reached the heavily forested island, we had to go ashore on small rowboats that would never have been accepted by any safety standards. As we worked our way through a dense forest following a narrow path, we came to a clearing where a gigantic banquet table covered with fresh vegetables, Russian dishes and wine and vodka awaited us. Speeches, awards and dancing to live accordion music soon followed. In spite of the language barriers, we all became great friends.

During a scientific meeting near Novosibirsk, Siberia, Yonas attends a lavish banquet on a heavily forested island.

The next day of the conference, Dmitri Ryutov, a leading Soviet plasma physicist at Budker's institute, had me, my colleague Paul Miller and Leonid Rudakov transported in a very old black limousine to a genuine Russian steam bath in a log cabin hidden in a Siberian forest. Our treatment included lying in a steam-filled room while attendants beat us with birch branches that had been soaked in boiling water. This was followed with a snack of small, hard bagels and dark, fermented liquid. Then we spent the remainder of the afternoon chopping wood and shooting skeet while discussing Russian culture. During the meeting we also talked about data and theory regarding high current electron beams. This was typical of the many meetings I attended – where the science was less important than the creation of a community of scientists from all over the world that cemented a lasting collaboration and friendship among all of us. Unlike many American politicians, administrators and military, I was building a personal bond with Soviet scientists and engineers. We had shared our meals, experiences and cultures, had met each other's families and had laughed, sang and drank together. A few years later, I would sit in meetings and listen to military leaders discuss the threat of the Soviet Union, but for me the Soviet citizens had become real people, not a nameless, faceless threat. Of course, as a naive and uneducated amateur in Soviet history, I knew nothing of the evil forces that had resulted in the unimaginable horrors that had been inflicted on the Soviet people.

I often found that the Soviet scientists could out-think us, particularly in mathematics, but could not out-spend us.

One of them commented to me that a combination of "our brains and your money could accomplish anything." In fact, the lack of money had become an issue for their entire country. During the 1970s, many of the country's citizens dealt with daily shortages of goods. On my first trip, I went to the largest department store in Moscow and found long lines and empty shelves. I recall a three-day visit to Leningrad in which the only food available was bread. I really loved the dark, moist Russian bread, but I imagined that the average Soviet citizen yearned for a better life. Historian Leonid Katzva remarked in his writings: "In order to buy furniture, a washing machine, a fridge, not to mention a car, one had to subscribe to a waiting list in a store, or more often at a plant, and be ready to wait for several months or even years. At the same time, if one somehow fell on the wrong side of the authorities, one could just as easily lose one's place on the list."[17] The Soviets were, however, proud of what they had achieved since the end of the "Great War" and tended to have a rather dim view of American society. They often spoke of their country's improving conditions and claimed, "It is not where you are in life, but where you are going … and how fast you are getting there." Years later, I would read the memoirs of Soviet insiders in the 1990s and realize what might have happened if more of the American decision-makers had had a better understanding of our chief competitor.

CHAPTER 6

The Soviet mindset

IN 1975, I INVITED A group of Soviet scientists to visit Sandia in conjunction with an international fusion conference, called the Beams Conference, which I was hosting. Part of the attraction would be a visit to Sandia facilities that had previously been closed to them. Because Sandia was located on an Air Force base, making the arrangements required a great deal of pleading, planning and negotiation before the Air Force officials grudgingly agreed to let "those #@%$ Rooskies on our base." I remember telling one of the invitees that he would be the first Soviet to visit Sandia and, with a raised eyebrow and the dry humor typical of his countrymen, he replied, "I will be the first – that you know of."

Although my guests may not have had ever actually visited the lab before the conference, I was certain they had studied satellite surveillance photos of the base, memorized our route to the lab and were prepared for the landmarks they would see. Little did they know that right before we picked the Soviets up from their hotel to transport them to our facility, the Air Force officials had asked me to cancel the visit due to classified activities taking place along the route to the lab. Quickly, I contacted the Sandia security people and they found a back road through the canyon at the outer boundary of the Air Force base. We went hurtling over an unpaved back road through a dry ravine filled with tumbleweeds and jackrabbits. It looked like a scene from an American western movie, and I halfway expected to see Indians on horseback riding toward us to attack our bus. The Soviet scientists couldn't hide their surprise. One of the other conference attendees was a colleague and close friend, Charlie Martin from Great Britain, who had visited our lab many times and knew the correct route to the facility. When he saw the dirt road and the shocked looks on the faces of the Soviet scientists, he could not help but laugh at what he thought was my clever trick.

British physicist Charlie Martin examines a piece of equipment during the 1975 electron beam conference at Sandia Labs in Albuquerque.

Despite this excitement, the conference went well and we continued to make visits to each other's labs. Rudakov and his colleagues frequently came to my home for dinner, bearing gifts of traditional Russian handicrafts for my wife and daughters. During one of these visits, I took Rudakov to the shopping mall to purchase American souvenirs such as Levi's to take home to his family in Moscow. The brilliant scientist found American capitalism completely confusing. "Why are there so many types of blue jeans at so many different prices?" he asked, shaking his head and leaving the stores empty-handed. I realized then that at least one of the profound differences between us and the Soviets was the foundation of their economy that had been dominated by the military. I often offered to bring items on my trips from America to my colleagues who were undeniably much better off than the typical Soviet citizens. One top Russian physicist from the Lebedev Institute once asked me to bring phonograph needles. In another conversation, he showed me his passport that clearly identified him as a Jew, and he hinted at the hidden level of anti-Semitism. Another requested a pound of coffee that we could buy at any supermarket. Gradually, I gained an understanding of the austerity of life in the Soviet Union and the underlying stresses of their closed society. My colleagues did not have some of the most ordinary consumer goods, but more important, they lacked our freedom.

Despite the complexities of souvenir shopping, Rudakov showed great enthusiasm for his travels. He appeared to love the mountain air and desert landscapes of Albuquerque, and I vividly remember the night he flung himself down on the lawn in our backyard so he could view the Milky Way shining brightly in the high-altitude air. Perhaps I would remember that occasion less vividly if I had remembered to clean up the yard before the visit. We had a large backyard – but we also had a large family dog.

Rudakov traveled with a man who appeared to be a "watcher" or "bodyguard." I surmised that this gentleman was a KGB agent there to ensure that no Soviet secrets were disclosed. My children couldn't keep up with the Russian names and soon began referring to Rudakov's watcher as "Off" and Rudakov as "Koff." During one visit, Off, the watcher, became a silent fixture at all our meals and meetings with Rudakov but never joined our conversations since – according to Rudakov – he knew no English. Indeed I never heard the apparent KGB agent utter a word of English, but at my home, my wife walked in on a clear conversation in English between my 3-year-old daughter and Off one night after dinner when the rest of us were out of the room.

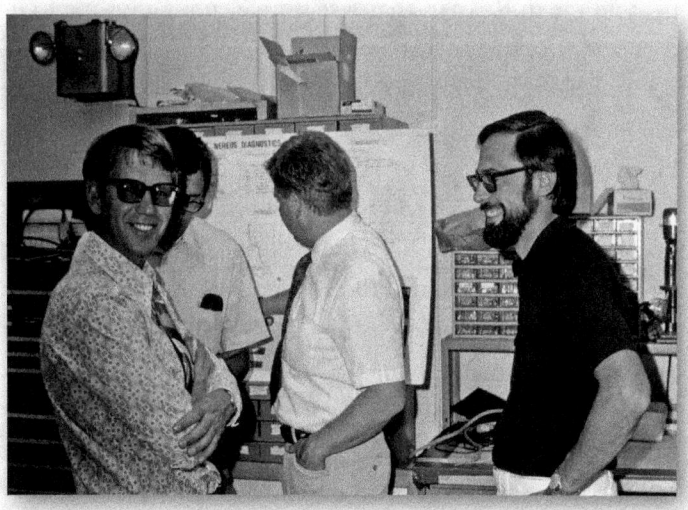

Soviet scientist Leonid Rudakov and his "watcher" talk with Sandia scientists during a visit to Albuquerque.

A few years later, at the end of one of our meetings in Leningrad, several Soviet scientists gave me copies of recent papers they wanted shared in the U.S. I did not consider the implications of this until I was suddenly stopped on my way through customs at the Leningrad airport. I was about to gain firsthand knowledge of the Soviet culture of secrecy and the protocol for protecting information not approved for release. A Soviet soldier took me to a barren office where, without any explanation, he had me wait for what seemed like an eternity. Finally, a very formal and intimidating looking colonel entered and demanded the details of my visit. I calmly said that I was an official visitor who had been invited by the Soviet Academy of Science and suggested that he call Kurchatov Institute Director Evgeny Velikhov to confirm. Then I noted that I didn't want to miss my flight. The colonel told me in Russian that was simple enough for me to understand, "Be silent." He ordered me to open my suitcase, and out tumbled the collection of papers. As the colonel studied the papers, I explained that they were not secret documents but scientific publications. He pointed out that they did not have the required stamp on the back that said they were for open publication, and he proceeded to study the papers one by one. When I reminded the colonel that he was making me miss my plane, he ordered me again to sit still and shut up and pointed out that the plane was long gone.

Now I was getting worried. The colonel began to methodically write out the title and author of every document followed by a few sentences in Russian. Then he asked me to write my signature at the bottom of the page. U.S. government officials had warned me never to sign any documents on such trips to the Soviet Union. Who knew what they were claiming – I could have been admitting that I was stealing Soviet secrets. Next stop – the gulag or even worse! But I also knew that I wanted out of that office, so I decided to treat the situation lightly. Fortunately, the colonel was simply following orders. In the Soviet Union bureaucracy took precedence, and the key was to do what you were told to do. "So colonel, if I sign this confession, can I take the documents out on the next plane?" I quipped. The colonel broke into a smile for the first time and nodded yes. I grabbed the pen and signed.

"I hope the visit in the Soviet Union was enjoyed," said the colonel in heavily accented, broken English, and he ordered the waiting Soviet officials to escort me back through customs and to the airline gate. To my surprise, my plane still was there and the other passengers had been waiting for what seemed like hours. They scowled at me with contempt for their delay. Finally, I was about to board when I heard hurried footsteps and someone yelling for me to stop. Out of the corner of my eye I saw what looked like an armed soldier rushing toward me. I considered sprinting onto the airplane, but suddenly the colonel who had become my "close friend" grabbed my arm and said, panting from chasing me to the airplane, "My apologies, here is your documents." He had forgotten to return one of the scientific papers that had caused all the trouble. I grabbed the document with relief, and the colonel saluted and walked away.

At the next international meeting in Paris, I told my Russian colleagues what had happened and warned them that they should not expect me to return to their country for a very long time. One of the scientists said he had heard about the incident and he told me not to worry, adding, with a smirk, "We know you are not a spy." Later that evening, at a crowded cocktail party, someone sneaked a paper bag beneath my arm and was gone before I could turn around. I rushed to the restroom to open the bag in private, and there I discovered a book titled *Secrets of Soviet Science* autographed by all my Russian scientist friends with the inscription, "From Russia with love."

Needless to say, I returned to Russia. On my next visit, when the plane landed in Moscow, the flight attendant escorted me out before the other passengers. As the door of the plane opened, I was surprised to see Evgeny Velikhov, the head of the Soviet nuclear program, waiting for me. Bypassing customs and security, Velikhov escorted me into a beautiful reception area and offered me tea, cookies and vodka. He then casually asked me if there was any truth to the rumor we had successfully produced fusion with our new machine. He was clearly worried that I would make some surprising announcement that would catch him off guard. "Of course not," I said, despite our promising experiments. Velikhov then introduced me to a man named Boris who was to be my escort at the meeting and told me to ask Boris for anything I might need. "This time," said Velikhov, "Russia will treat you how it should – with open arms."

Throughout the meeting Boris was on hand to assist me, and I received VIP treatment from everyone involved. During a break between presentations, Boris introduced me to a reporter from the Soviet newspaper *Izvestia* who asked if I would be willing to provide him with an interview about the conference. I agreed. The interview questions quickly strayed from the issue of fusion research to questions about fusion weapons, so I stopped the discussion. I later told Boris what had happened, and Boris apologized. The next day Boris approached me and whispered into my ear, "Professor Yonas, the reporter no longer exists." Was he fired? Sent to the gulag? Or was this just an example of Soviet humor? I never did find out exactly what it meant that the reporter no longer existed and – in many ways – I didn't really want to know. Again, the Soviet emphasis on secrecy, bureaucracy and protocol was evident, and my American understanding of their culture was limited at best.

CHAPTER 7

Soviet laser technology

———

THE EXTREME CULTURAL DIFFERENCES BETWEEN the U.S. and the USSR influenced far more than my relationship with my Soviet colleagues. These differences impacted every aspect of interaction between the two nations and every decision the political leaders made. At the heart of the superpowers' inability to understand one another lay a sense of distrust and hostility that began in 1922 with the formation of the Union of Soviet Socialist Republics under communist control. Profound ideological differences, the legacy of Josef Stalin, the Soviet expansion in Eastern Europe and the American desire to stop the spread of communism all combined to create an atmosphere of unresolvable conflict. Helping fuel this conflict were technological developments in offensive and defensive weapons on both sides.

Ever since the realization that intercontinental ballistic weapons carrying nuclear weapons could destroy cities from great distances in less than one hour, the Soviets became committed to developing a defense against such weapons – along with offensive weapons that could allow them to mount a retaliatory strike. In 1957, the Soviets launched Sputnik and

America reacted by creating the first intercontinental ballistic missiles. In response, the Soviet military leaders turned to their scientific community to bring them a new and better way to protect Moscow from nuclear-tipped ballistic missiles.

Leading one segment of the Soviet weapons research programs were two very different kinds of scientists. N.G. Basov was formal and humorless – a theorist who focused on the fundamentals. Basov, who had received the Nobel Prize along with Prokorov, another key Soviet scientist, and Charles Townes from Bell Labs, for the invention of the laser, proposed a nuclear explosion to create a 10 million-joule laser pulse (with an energy equivalent to six sticks of dynamite). At that time the largest existing laser was just 10 joules but Basov's colleague, technical genius Oleg Krohkin, an optimistic inventor, helped fuel speculation that they could create a 10 million-joule laser pulse. Another leading Soviet scientist, Evengy Velikhov, who became a colleague of Leonid Rudakov in the quest for fusion, was a gregarious engineer with a practical, businesslike approach to getting things done. Velikhov also set his sights on creating a 10 million-joule laser pulse, but he aimed to create a continuous rather than short pulse using a magnetohydrodynamic (MHD) power generation instead of a nuclear explosion. The Soviet scientific leaders and their staffs were competing against the Americans – and each other – to achieve a scientific breakthrough in fusion, weapons and missile defense. Basov soon switched his concept to a more practical chemical explosive-pumped iodine laser and achieved a 1 million-joule (MJ) output. This demonstration launched the Soviet Union's Terra 3 program and led to the creation of many rapidly built giant Soviet facilities, sparking great concern and confusion within the U.S. From our satellite imagery we knew something big was happening, but we did not know what it was.

The Terra-3 laser scientific development program was located at the Soviet nuclear weapon laboratory, Arzamas, and the testing center was headquartered at Sary Shagan, an antiballistic missile testing range in Kazakhstan. The construction at Sary Shagan began rapidly as the Americans watched from satellites, and the facility became a vexing enigma for the U.S. intelligence community. Meanwhile, the explosively pumped laser was rapidly scaled first to 10 thousand joules and then, in 1969, to the absolutely stunning goal of one MJ. The Soviet Union had made this achievement in just three years instead of decades, and construction of the facility and implementation of a real anti-ballistic missile weapon program had begun. By 1973, the Soviets were building giant lasers and by 1978 they were testing the laser against missiles, airplanes and other military equipment. The Soviet scientists could taste success, and the military's enthusiasm provided all the needed rubles to move rapidly to weapon applications.[18]

As it turned out, the angel of early success morphed into a devil, and the Soviet researchers discovered they had a giant flashlight rather than a giant laser. The beam diverged rapidly and could not be focused onto a distant target. They could only destroy a target at close range. Instead of building a destructive laser weapon, they had created more of a powerful flame thrower. From the beginning, the Soviet scientists knew that the beam would need to have a small divergence in order to be a practical weapon, but they had convinced themselves that the divergence problem could be solved. As in many such early successes that then lead to another invention that leads to yet more inventions, a research program can grow in size, in funding and complexity, without hope of really achieving the goal. As money is spent, promises are made, people are hired to work at the facilities, program reviews are given to visiting military decision-makers, awards and medals are distributed, and it is easy to see that self-deception could emerge – particularly with the military looking over every move made by the scientists. Soon the program was out of financial control and had grown into a giant community of engineers and scientists with competing ideas. This type of spiraling spending and scientific competition was certainly not limited to the Soviet Union. A similar behavior was emerging in the U.S.

The Sary Shagan facility had been a failure. Basov shifted his attention to focus on fusion, which also would need a one- to 10 million-joule laser pulse, but with a duration of a few billions of a second. He later rationalized the time and money devoted to the Terra-3 program by saying, "Well, we made sure that nobody can shoot down a ballistic missile warhead by a laser beam."[19]

Peter Zarubin, a Soviet industry official and partner in the development program, also rationalized the benefits of the expensive program. In a review of the program published in 1995, he wrote, "These achievements serve as the foundation for the development of high power laser technology in Russia even today. ..... Many scientific and technical achievements were used in many subsequent research, including inertial thermonuclear fusion .... many thousands of scientists, engineers, and military and administrators have made important contributions ... over 10,000 relevant publications."[20]

CHAPTER 8

The view from the U.S.

While many Soviet scientists saw Terra-3 as a gigantic failure, the program quickly captured the attention of military intelligence experts in the United States. Just as they were nearing the end of this expensive and ultimately discouraging investment, the Soviets were stunned to learn that the highly decorated Major General George Keegan, the head of U.S. Air Force Intelligence, had announced his discovery that the Soviet Union had not been working on a laser weapon as originally suspected. Instead Keegan announced that he had determined that a ground-based particle beam weapon had been developed in Semipalatinsk, Kazakhstan. The Soviet weapon, said Keegan, was "capable of destroying the entire USA capability within one strike!"[21]

The Americans had dubbed the Soviet facility PNUTS, which stood for a Possible Nuclear Test Site. That was not particularly insightful since it was located in the middle of the Soviet nuclear weapon test range. This facility was indeed such an enigma to the U.S. intelligence community that the

CIA asked one of the remote viewing experts in their Stargate paranormal program to take a look ... so to speak.[22] This viewer was a retired Los Angeles police department sergeant with the alleged ability to "see" distant sites psychically. After he was given the geolocation of the Soviet Terra-3 facility, the sergeant drew an image of a giant A-frame crane on rails. His "clairvoyant prowess" clearly revealed something was going on at the facility – but no one in the U.S. knew what it was or what it meant. I was one of several American scientists the military brought in to try to decipher these mysterious paranormal drawings. All I could do was speculate that the Soviets were building "something" beneath that giant A-frame. Clearly, my psychic abilities were limited at best. Incidentally, a friend of mine later visited the facilities and brought back a photograph of the giant crane that was used to cover the opening to the underground facility when satellites were overhead.

General Keegan insisted that the Soviet threat was imminent. In May 1977, the American national security community was startled by an article in *Aviation Week & Space Technology* in which Keegan discussed Soviet progress in directed energy weapons. He was quoted as saying, "In increasing numbers, U.S. officials are coming to a conclusion that a decisive turn in the balance of strategic power is in the making, which could tip that balance heavily in the Soviets' favor through charged particle beam development ... Most of the controversy centers on what tests are being conducted in an unusual research facility about 35 miles south of the city of Semipalatinsk."[23]

Aviation Week editorialized, "The Soviet Union has achieved a technical breakthrough in high-energy physics applications that may soon provide it with a directed-energy beam weapon capable of neutralizing the entire United States ballistic missile force and checkmating this country's strategic doctrine ... The race to perfect directed-energy weapons is a reality. Despite initial skepticism, the U.S. scientific community is now pressuring for accelerated efforts in this area."[24] Clarence A. Robinson, a writer for the widely read magazine, then announced that the particle beam weapon breakthrough in ballistic missile defense had already been tested at Semipalatinsk and that satellite data proved his theory. Fortunately, *Aviation Week*, which was often sarcastically called Aviation "Leak" by U.S. aerospace engineers, could not have been further from the truth.

In actuality, Semipalatinsk was the center of the Soviet nuclear rocket program. Keegan's interpretation, based on the satellite images, was so profoundly wrong that it allowed for us to miss the real Soviet program. His interpretation also even affected the Russians by ironically boosting the budget for the Soviet research program and inspiring Soviet scientists. One Soviet commented later that "... many young Russian scientists in the 1980s were thrilled to be sent to Semipalatinsk, where they assumed they would be working on 'Keegan's beam' and when they arrived they would be greatly disappointed as they realized that the mentioned laser program did not even exist!"[25]

CHAPTER 9

Rudakov's real Soviet breakthrough

KEEGAN BASED HIS ERRONEOUS CONCLUSIONS partially on an actual Soviet breakthrough which was achieved by my colleague Rudakov at the Kurchatov Institute. This discovery was as significant and dramatic as the circumstances surrounding the way Rudakov revealed his findings to the international scientific community.

I first learned about Rudakov's experiment at a Gordon Conference in Santa Barbara, California, in July 1976. These meetings are for sharing of unpublished scientific data and ideas at an early stage in development that are not ready to be shared publicly. The rules of the Gordon Conference are like the Las Vegas rule – what happens at the Gordon Conference stays at the Gordon Conference. This particular Gordon Conference would reveal a scientific breakthrough that would be very difficult to keep under wraps.

I was chairing the evening session at the conference in which Rudakov was supposed to speak. Earlier that day, he asked me to meet him so he could tell me something important. He said it was something he wanted to share with me first, as a friend and a colleague, before he revealed it at the evening session. Following Rudakov's instructions, I met him on a lonely beach that day at noon. No one was around. This seemed very mysterious to me, and I wondered if Rudakov could be in some kind of trouble. He beckoned me to kneel beside him on the abandoned beach, so, despite feeling as though I had become a star in a low-budget spy movie, I hunkered down at his side.

Rudakov picked up a stick and began drawing in the sand on the beach. His drawing revealed a significant Soviet breakthrough in top secret fusion research. The Soviets had discovered the concept behind the H-bomb. Although the U.S. had already discovered this so-called Teller-Ulam H-bomb concept, we did not know the Soviets had achieved the same breakthrough. The idea of using X-rays to drive the fusion capsule was not new to us in the weapons program, but it certainly was not for open discussion. Rudakov explained that he had successfully converted the electron beam into X-rays, and the X-rays then drove the implosion to produce a small fusion reaction. He then dragged his hand over the drawing in the sand, erasing the evidence of his revelation, while I sat, mouth open, on the empty beach in shock.

That night, Rudakov gave his presentation, sparking an immediate reaction from U.S. security officials who advised the Americans in the audience that the talk was classified and seized the blackboard on which Rudakov had written diagrams and equations during his presentation. According to a 1977 newspaper article, "mouths dropped open all over the room" at Rudakov's revelations and security officials sent telegrams to four U.S. nuclear research laboratories "advising them to 'play dumb' if asked about Rudakov's talk."[26] As for Rudakov, he claimed his disclosures had been reviewed and approved for release by Soviet security. Some U.S. officials speculated that the Soviets deliberately revealed the breakthrough in order to warn America of their advancements and prompt the U.S. to agree to a strategic arms limitation agreement.[27] Personally, I wondered whether it was an intentional leak or whether the Soviet security system had simply screwed up and failed to read over Rudakov's proposed presentation paper. Meanwhile, Keegan used Rudakov's disclosures to support his belief that the Soviet particle beam weapon program was successful and to push for increased U.S. spending on research to outpace the Soviet threat.

Looking back on this incident, I realize that the implications for the Department of Energy security apparatus were profound. By trying to keep Rudakov's revelation under wraps, the DOE left plenty of opportunities for the news media to make up wild stories. One of the strangest claims was that "the Soviets are nearing a breakthrough in developing nuclear weapons 100 times more powerful than the largest current weapons .... a gigaton hydrogen bomb ... a doomsday bomb ... destroy the world in one blow."[28] Meanwhile, the physics community widely discussed attempts to declassify the concept even though Rudakov's "blackboard notations were confiscated by the FBI seizing the blackboard."[29] Two of the most prominent physicists at the time, Ray Kidder and Hans Bethe, argued over whether to officially acknowledge Rudakov's breakthrough. Bethe was concerned that this information would somehow lead to proliferation of the secret of the H-bomb and "serve them this idea on a platter ..... a very damaging act ... an easy road to H-bombs can only stimulate them more on the road to A-bombs and then H-bombs."[30] Kidder concluded, "It seems likely that the secret in question will fail to meet commonly accepted standards of what constitutes a secret, in view of the existing published literature"[31] Eventually, the science community accepted the fact that there was no way to hide Rudakov's disclosures, which were finally accepted for open discussion. The intersection between the communication of complex technical issues and the need to maintain security would soon rise again as the debates over missile defense increased after Reagan's Star Wars speech.

CHAPTER 10

The Star Wars speech

───────

BEFORE I WAS ASKED TO help manage one of the largest research and development programs in the history of the U.S., I and many others had been given the opportunity to provide advice that should have prevented this entire adventure in government spending. President Reagan had decided – largely on his own – that our nation should embark on a program to develop a defense against ballistic missiles without consulting the scores of technical experts who had been studying this subject for decades. He had been given lots of advice from strong advocates from the military, industrial and congressional sectors that stood on the right wing of the political spectrum, all with their own notions of various yet-to-be-developed technologies. These primarily political ideological advocates were balanced by many more experts who were on the left of the political spectrum and tended to be members of the academic community.

Reagan had opportunities to engage the knowledgeable technical experts. He had created his own commission to advise him on our strategic deterrent that was in disarray because of the perceived vulnerability of the yet-to-be-deployed MX missile. This strategic weapons commission, which consisted of many experienced experts, was expressly opposed to any sort of missile defense. A few years before that, I had served on a committee to advise the Pentagon on the subject and had come up with many suggestions regarding how to spend vast sums to find a dimly lit path that might conceivably lead to a solution to our strategic deterrence problem, but the Pentagon and the Congress were in no mood to invest in such a high-risk quest.

Despite this dearth of real support for a grand new program, March 23, 1983, in a televised address on national security, President Ronald Reagan delivered the historic speech that led to the creation of the Strategic Defense Initiative. The nation watched transfixed as Reagan exclaimed, "What if free people could live secure in the knowledge that their security did not rest upon the threat of instant U.S. retaliation to deter a Soviet attack; that we could intercept and destroy strategic ballistic missiles before they reached our own soil or that of our allies? … I call upon the scientific community in our country, those who gave us nuclear weapons, to turn their great talents now to the cause of mankind and world peace, to give us the means of rendering these nuclear weapons impotent and obsolete."

When the speech was over, the room in the White House was overcome by silence. The reaction was immediate. The last words had been barely finished when two camps had been created. The side that called it visionary came from the aerospace industry and military research labs. Those who called it pure delusions and deception, including leading scientists such as a Nobel Prize winner in physics, Hans Bethe, saw it as a dangerous escalation of the arms race. Although most major newspapers almost overlooked the story at first, the next morning's *New York Post* ran the gigantic front page headline: "Star Wars Plan to Zap Red Nukes." The unofficial name of the program had been born.

The announcement of SDI took the world by surprise. Was Reagan clever, delusional or just acting? Reagan's understanding of the crazy acting method was instinctive. He knew that everybody was afraid of playing chicken with a mad man. Could the SDI be a part of that, or did Reagan indeed think the technology was possible? Was there a new breakthrough? Was this a new arms race? Or was it just a gambit to send the Soviets into bankruptcy? Reagan had a practical understanding of the need to keep new ideas very close to the vest, even from those near him. He understood the need to prevent people from dissuading him with doubts and questions. He didn't understand or even think about the truth of the science. This was about a vision, about changing the world.

In "Way Out There in the Blue: Reagan, Star Wars and the End of the Cold War," excerpted in *The Times*, Frances Fitzgerald says the SDI reflected the actor in Mr. Reagan. "A perfect anti-ballistic missile defense was beyond the reach of technology. It was just a story, and yet to trust the polls, the idea had great popular appeal in the mid-80s, and many Americans believed such a thing could be built. In that sense the Strategic Defense Initiative was Reagan's greatest triumph as an actor-storyteller."[32] Reagan's ability to maintain secrecy increased the enigma of the whole concept and made it almost impossible for friends or enemies alike to fully ascertain what this *thing* was, and to know just what to do with it, how to embrace it or attack it. Even though I was openly leading Sandia's fusion program, I had been working for more than 15 years on secret subjects related to nuclear weapons and "death rays." I had been interacting with many scientists and engineers throughout the world on many of the underlying technologies and published and presented papers on related subjects. Since I had not uncovered any practical solutions to the many scientific and technical challenges of ballistic missile defense, I interpreted his statement as an example of political rhetoric rather than a real policy. I did not know what was going on inside the administration, nor did I realize that a presidential directive was signed two days later to make a major shift toward defense and away from offense.

CHAPTER 11

The Soviet reaction to the Star Wars speech

WHEN RESULTS OF ANALYSIS AND defense implications reached the Soviet Union, the Soviet leader, Yuri Andropov, and his top brass were filled with confusion and anxiety. It suddenly changed the way they listened to the evaluations of their own experts regarding whether missile defense was indeed a possibility or mere fantasy. The mere "what if" was sufficient to cause fear. In the words of a friend and Army general on the Joint Staff, Reagan's speech "made the Sovs go ape shit."[33]

When their top scientists proceeded to express their well-informed doubts about the science behind it all, the generals would storm back at them with questions such as, "But what if there is something you don't know? What if they have discovered something we are not aware of? Why would they risk ridicule by coming on with such sense of certainty?! Once they put their mind to something, they somehow make it happen! We cannot let that happen again!"

The head of their nuclear energy program and one of the most influential scientists in the Soviet Union, Evgeny Velikhov, knew better. Velikhov, with his knowledge of the underlying science and technology, must have argued with the generals: "Let them try. Let them spend their billions on something that will not work. Let them bankrupt themselves!"

The argument probably continued: "What we need is to increase our own program. What you need, professor, is more funding and more development so we can compete!" Velikhov did what scientists I have known never seem to do. He did not respond with a request for more money. Instead, he continued steadfast in his resolve as if saying, "We cannot compete, nor should we." Velikhov must have been very frustrated, and I am sure his calm and scientific response was, "Shouldn't we first even find out more about what that program is before we react?"

Three days later, Andropov announced his reaction at a meeting of the Warsaw Pact. I can imagine him saying, "This is a plot to militarize space. SDI is nothing else but an effort to bypass all existing treaties and attack from the skies. We will continue to actively pursue our long established path of peace, and seek further treaties to outlaw any tests or deployments of space weapons. We cannot allow for U.S. military superiority, and we will not allow it. We will pursue any and all methods necessary." On March 27, Andropov declared in *Pravda*, "This is a bid to disarm the Soviet Union in the face of the American nuclear threat."[34]

SDI became almost immediately the largest focus of the Soviet's diplomacy and covert action. Their attitude toward SDI did not change upon the death of Yuri Andropov and rise to power of Valentine Chernenko as secretary-general. Why was SDI of such concern for the Soviets, even when it was merely in its forming stages and there was nothing but unsubstantiated claims?

A KGB analyst said it succinctly years later: "The whole SDI was more than anything a powerful psychological weapon. It emphasized our technological backwardness and told us to our face that we should re-evaluate our place in the world's technological progress." Most people thought SDI was about ballistic defense, but in my opinion, the reality was that this was only a side issue. Ironically, at a time where the Soviets were convinced we could resolve anything with our miniature electronics and giant computers, our scientific community knew we did not have a clue on how to make the necessary programs work. Robert Gates, several years later when he was head of the CIA, stated, "I think it was the idea of SDI and what it represented which frightened them. It showcased the emergence of American economy and technology and the slow demise of their own. There was great faith in the wake of the Manhattan and Apollo projects that with enough money and effort, U.S. scientists could overcome any obstacle." As it turned out, I was one of those U.S. scientists, and I did not have a clue.

CHAPTER 12

The Fletcher Study

On May 24, 1983, I received a call from Harold Agnew – a scientist who had participated in the Manhattan Project to develop the atom bomb and had flown on the chase plane during the Hiroshima bombing in 1945. Agnew had been the head of Los Alamos Lab, which developed most of our nuclear stockpile, and was a frequent advisor to the government on all things nuclear. He was very well connected and highly respected in all the science/political circles, but he was also known as an incurable kidder, and I liked that side of him. I knew him mostly from occasional meetings, and we had a good relationship because of our shared sense of humor. He could be serious and funny at the same time, such as when he proposed that all national decision-makers should observe nuclear explosions without a stitch of clothing on so they could feel the heat. He liked to poke fun at the establishment and the other national labs.

Agnew told me that the plan for the president's SDI program had to be delivered in only a few months, and he asked me to put together and lead a panel on directed energy weapons (DEW) that would be combined with the work of four other panels in a report that had to be finished in August. At first I doubted that he was serious, but the story had enough truth in it that I had to learn more. The head of the study was Jim Fletcher, former head of NASA. Agnew was helping to arrange the panels, and he would be the second in command in the study. He said the study would start in one week in two rented floors in an office building in Arlington, Virginia. Agnew instructed me to assemble a small group from labs and the government agencies rather than from the aerospace industry because those companies would be involved in any large production programs that might follow. Our salaries would have to be paid by our parent labs and government organizations, and there was no time to work out new contracts. It was obvious that this would be a full-time job for three months, but since the customer for our work was the president of the United States, there was no question about whether to get on with it. I had one week to assemble a team, temporarily assign one of my colleagues to run my fusion research program at Sandia, find a place to live in Washington and try to figure out what was really going on. After I decided that Agnew was serious, I then had to convince my wife that this was worth pursuing even though it meant I would desert the family for three months. Although at first Jane thought this was another of my grand jokes, she went along with the plan because it sounded like an important assignment that might lead somewhere. She also knew I was attracted by the excitement and challenge that was becoming less frequent in my real job.

Before we even had our first meeting in Washington, it was obvious that there would be no problem finding volunteers to participate, but the bickering was soon underway about who would be chosen. I had known all of the "likely suspects" for several years and had been involved with them in a mini-directed energy weapon, or death ray, program plan development with most of them in 1979. The 1979 plan had included practically all the U.S. participants in DEW research, and I already had my list of experts to cover all the bases: lasers, particle beams, nuclear-driven DEW and microwaves. This 1979 plan had been carried out for the research and development office of the Pentagon, and there were 50 of us who had worked for three days in a conference at Los Alamos to deliver it. All of us involved in the plan had a vested interest in its success, and there was no question about whether there were any conflicts of interest. This could be best described as the pigs designing the trough, or as we sometimes called our study, "Death Rays for Dollars." After the report was delivered to Assistant Secretary Ruth Davis in 1979, nothing happened, because, as I found out, there was no real appetite to spend more money on research and development, particularly for the "far-fetched" ideas we had generated; but, thanks to Ronald Reagan, things were about to change.

As the preparations for the Fletcher study were beginning, I found out immediately that I needed to choose the politically right person at LLNL. Most of the people at LLNL were deeply committed to nuclear weapons, but I did not anticipate a major role for nuclear weapons, so I chose Richard Briggs, who I knew and respected, to represent the particle beam approaches. This caused a dust-up that was blamed on Agnew, who was known as more of an honest critic about missile defense technology rather than an enthusiast. Officials at LLNL told me that Agnew was "undermining the entire process, and nothing would come out of it if I persisted."[35] Of course, Agnew then told me to do what I thought was right even in the face of threats to sabotage the study. He told me to ignore all pressure from special interests and when the going got rough, to just laugh it off. He believed that humor was a lot better than worrying.

The entire process was starting off poorly with lots of conflicts, confusion and hard feelings, but Agnew just brushed it off as his normal way of dealing with this kind of "high-level mumbo jumbo." I quickly learned not to take such vitally important things too seriously, particularly when Agnew was involved. One day, however, Agnew became surprisingly serious. He took me aside and warned me in a very grave tone of voice that I needed to very careful. He said the process could be really dangerous and lives would be lost. I was shocked at this suggestion, but then Agnew raised an eyebrow and continued by saying the situation could quickly turn deadly because "there would be so many contractors trampling each other on the way to the sources of funding."[36]

I should have seen Agnew's quip coming, because I had experienced his sarcastic streak before. One day when he was testifying to the Joint Congressional Committee on Atomic Energy, the senator chairing the session told him to cut his remarks short. This was not a surprising request from the harassed congressmen who seemed to always be in a hurry to be somewhere else. In an instant, Agnew pulled out a huge scissors from his pocket and cut his viewgraphs in half. The general who ran the nuclear weapons program was furious with Agnew's disrespectful behavior, but I thought it was a brilliant gag, and I always wondered why he was carrying the scissors.

In spite of Agnew's lighthearted attitude, I knew this was totally serious business and I was determined to behave myself. I thought I knew how to proceed in a disciplined manner by collecting, sorting and prioritizing proposals into a budget proposal, but I still did not really understand the overall goal of the study. On June 1, Fletcher and the key panel chairmen who were to be the study directors got together at 6:30 p.m. in a small conference room to prepare for the first formal briefings. In a very tense meeting we learned that there would be two separate studies, one on technology and one on policy. This was my first inkling that the engineers and scientists were not going to have a clear definition of the strategic goal.

In spite of the importance of this event, everything started off with what seemed like a big muddle of confusion. One thing Fletcher made clear to all of us was that this was the president's program and we were to come up with both the optimum program "whatever the cost" along with a fall-back option with a stretched out schedule if we were limited for funds. The implication, however, was that we would get whatever funding was needed. Agnew said from the beginning that "there may not be a pony in this pile of horse shit," and it was obvious from his sarcastic comments and generally negative view of all of this that he was going to be an honest skeptic no matter what we did. Nevertheless, there was a faint smell that could be sensed in that small room, not of horse shit, but of unlimited money. We all knew that Agnew was going to be a major influence on the final outcome of our study, and I wondered if he would remind us to ignore the smell of money and be more aware of the smell of manure.

We were to focus on ballistic missile threats and how long it might take to even know whether we could find Harold's "pony." We were under the gun to come up with an assessment including critical issues in only six weeks and a comprehensive long term R&D plan by Aug. 31. We knew we had to review all the prior reviews and the reviews of those reviews, give a fair audience to any reasonable proposals and deliver a product on time. The next day was "show time," and all the key administration representatives were to show up and get everybody off to a flying start.

The day started off in the large, main conference room filled with obviously newly rented tables and chairs that had been hastily acquired. The room was surrounded by smaller, rather cramped meeting rooms where the panels were to do their work. The room was jam-packed not just with the

50-odd members of the so-called Defense Technology Study Team (DTST), but also lots of hangers-on from the military, the administration and various agencies. The feeling of anticipation and curiosity was complemented by the discomfort of cramming too many people into a room that was inadequately air-conditioned. Because it was obvious that nobody knew what to expect, imagination ran wild that something big would come from all of this.

When I showed up early to take a look around, I was told the approach would be that the panels would work separately in our smaller rooms collecting, analyzing and summarizing vast amounts of information. Everything had to be done right there, with no excursions that might lead to leaks of information. Once a week, each panel would report back to the entire team in the main conference room so Fletcher and his executive team could try to make all of this into a coherent picture. The separate policy team was on a different floor and could have been in another world for all the guidance they were to give us.

We had plenty of security and control of all documents, and if you were not on the cleared and approved list, you did not get in. We expected that whatever we learned and all our work would be very closely held and not made public. I felt privileged and honored that I would be getting the inside story about this tremendously important program. I was to be terribly disappointed by what was about to transpire.

The first speaker was the study director, Jim Fletcher. Fletcher, the epitome of a dedicated public servant, was about as straightforward and credible as any technical leader I had ever met. You could believe everything he said in his low-key way, and he had no hidden agendas. He had been drafted for this job, just like the rest of us, and he probably did not appreciate the sarcasm and humor of Agnew, but he never showed any irritation or resentment. Fletcher made it clear that we were to focus on "protection of our people and our allies against all forms of ballistic missiles, including a full-out attack."[37] To me, he appeared to be laying out an impossible goal. At this point, I could not help wondering why we should even think about such an unrealistic goal and whether others could conceivably accept anything we might recommend.

Other administration representatives then focused on the glue of the program, namely the sensors and the battle management. It was clear that killing targets was only a part of the problem. We had to find the attackers in their half-hour trip from the Soviet Union in the vast expanse of space over the top of the world and figure out what to do with them, make a prompt and very precise decision to do it, and do all that in minutes before we even tried to intercept them. The press had made it sound like the hard part was the death rays, but it sounded to me that this could be the easy part. Precision detection and precision kill would most of all demand a precision decision – in split seconds. I wondered who would make this decision, or would there even be time for any person, such as the president, to be involved. If the decision was to be made by a computer, how could we ever trust it?

The next speaker was Jay Keyworth, the president's science advisor, and I figured he would make everything clear, but that was not to be. He gave us no numbers or concrete goals other than that we had to "protect the credibility of the president," since this whole thing was the president's idea. What the president wanted was a more stable world. Keyworth said the MAD doctrine was no longer going to be the basis of our strategy and was to be replaced by a defense strategy. We were not to consider protecting our deterrent forces, and we were to harness the power of our scientific and technical capabilities to change only everything that was the basis of our present-day strategic balance. He asked for "boldness, risk-taking, innovation and imagination."[38] Keyworth's presentation was long on inspiration but short on specifics. He told us that we needed to come up with near-term demonstrations and real deliverables ... of something that was yet to be determined.

This was the first time that I was faced with the dilemma of wondering "if it would work" without knowing "what *it* actually was." Fletcher had said it was to protect the people; Keyworth had said it was to protect the credibility of the president; and I had begun to wonder, "How did I get myself into this mess?" I then thought about the support from the president and I saw all the assembled talent in the room. I hoped we would figure something out. I had a great team representing the leadership in the nation's DEW programs, and they were all smart and dedicated to the science and technology. I judged them all to be people of high integrity who really wanted to make something happen. Surely with all this talent working for three months, we could come up with something reasonable and useful ... I hoped.

The dilemma of discovering a reasonable approach was finding a way to deal with the extremists. We knew there were strong advocates for each of the proposed approaches to death rays and there was enormous pressure from the advocates to "stop wasting time and get on with it." Many reviews, however, had already pointed out the obvious deficiencies of all the approaches. The strongest advocacy was coming from Senator Malcolm Wallop (R-Wyo.) for near-term deployment of chemical lasers in space, but the White House Science Council had just reviewed all DEW concepts, with a focus on the chemical laser status, and gave it a grade somewhere between a C-minus and a D. There were other emerging concepts that were not as well evaluated, so there was hope, but my panel had to figure out what was just a vision and what could be included in a real R&D plan. The other concepts got a grade from the White House Science Council of "incomplete," but they were willing to support the investment of some more money, maybe a few hundred million dollars, to answer many of the outstanding technical questions. It would not be an overstatement to describe their attitude as open-minded, but not very enthusiastic.

From the beginning, I saw the effort as a combination of policy and technology and I was struggling to understand the strategic basis for our work. At this point I was pleased that the undersecretary for policy, Fred Ikle, would be the next speaker. He was a highly respected expert in military strategy and policy, and I figured he could give us a more clear perspective. Unfortunately, he just made it all even more complex. He added another dimension when he told us to avoid messing up our alliances, and in particular not to offend NATO, which had not been consulted before the March 23 speech. Ikle offered the possible use of arms control agreements to restrain any Soviet response, and that seemed reasonable to me, but I wondered if it would be reasonable to the Soviets. I thought naively that maybe the Soviets would cooperate with us to help us achieve our goals if we could offer them something they really wanted. That would mean we would have to better understand their thinking, and I had little hope that any of us really knew how to think like they did.

Other speakers tended to muddle up the process with talk of easier to achieve intermediate deployment such as anti-satellite weapons. One speaker discussed the need to satisfy the impossible goal of one hundred percent effective defense, and another described already rejected improvements in deterrence by defending our missile silos that were thought to be vulnerable to the newly emerging Soviet missile developments. This approach was totally contrary to the president's goal of eliminating the threat to people. How could the administration provide this study with such confusing and contradictory guidance? Was the right hand talking to the left hand? Obviously not. I then knew I would not figure out the strategy, because there was none, and I had to focus on getting the technical facts straight. I decided we would listen to any reasonable proposals, but do our own analysis and get on with writing our report. Maybe the strategy would emerge from the work of the policy wonks that unfortunately seemed to be a bit scarce.

I was working full time against a tough deadline, with a poorly defined goal, and with lots of political and strategic implications of whatever we decided. After dozens of briefings, long days of arguing and even more days writing drafts and trying to get them approved by the rest of the committee, we were left with no clear recommendation that could be a response to the president's request. The tension was building, and in spite of my pledge to myself to maintain discipline, we needed some comic relief.

There had already been sufficient leaks of our DEW working papers that made their way, sometimes with typos and mistakes, into *Aviation Week*, which created concern that others, probably somebody at the White House, was trying to game the system. We were following strict security approaches with our notes turned in every night, blackboards erased and trash cans emptied .... so I figured we could scam the thief, whoever that was, by planting some made-up information. I invented a fictitious weapon, the Pluton Beam, as a guaranteed route to a robust ballistic missile defense. My panel put together a serious sounding combination of techno-babble that we figured the thief would accept and publish. We wrote about our advanced ballistic missile defense "based on the Josephson effect as described in the work of

Hansburry, Brown and Twiss leading to a photoelastic effect in materials with near-zero conductance."[39] With help from the security folks, we left descriptions in all the prohibited places in our meeting room, but unfortunately nobody took the bait. Eventually, when I left the SDI, I was given a T-shirt emblazoned with the emblem, "Father of the Pluton Beam" as a measure of the respect and confidence my team saw in my work and leadership skills.

When we were finished in August, there were opinions that ranged from totally negative to somewhat positive. The majority opinion was somewhere in the middle ground and, although we had a plan and a budget, we did not have a quotable conclusion. I told Jim Fletcher I would try to write the conclusion and try to get the other panel chairs to agree, so I wrote the following:

> We concluded that a robust BMD system can be made to work eventually. The ultimate effectiveness, complexity, and degree of technical risk in this system will depend not only on the technology itself, but also on the extent to which the Soviet Union either agrees to mutual defense arrangements and offense limitations, or embarks on new and more desirable strategic directions in response to our initiatives. Since the outcome of this initiation of an evolutionary shift in our strategic direction will hinge on as yet unresolved policy as well as technical issues, we recommend that steps toward early development

of systems of marginal capability be scrupulously avoided. ... No definitive predictions of the outcome can be made, but we must start the process which can lead us in this new direction.[40]

I had intentionally ended this statement with a very negative conclusion that "no definitive predictions of the outcome can be made" in order to emphasize that the outcome was not in our hands alone, but my real conclusion never saw the light of day. What finally made it into the well-publicized and short version of the conclusion was the following: "By taking an optimistic view of newly emerging technologies, we concluded that a robust BMD system can be made to work eventually."[41] The words *optimistic* and *eventually* told the real story, but the need for the Soviet Union to cooperate was not made clear. I then went back to Sandia and immediately dove back into the day-to-day work of the fusion program. I concluded that this entire adventure was over. Naturally, I was wrong.

CHAPTER 13

The Scowcroft Commission

In 1983, Reagan empaneled the Commission on Strategic Forces, chaired by retired Lieutenant General Brent Scowcroft. The commission's charge was to examine the future of America's intercontinental ballistic missile force.[42] In December of that year, Scowcroft was preparing to brief Bud McFarland, the president's national security advisor, on the commission's view of the Fletcher study recommendations. He asked General Bob Rankine, representing the Pentagon; Larry Gershwin, representing the intelligence community; and me, representing the Fletcher study, to share our plans with a sub-group of the commission. For me, this turned out to be a slap in the face with reality.

We were meeting in a small room in the old executive office building next to the White House, and the session started off poorly for me since I wanted to show a movie. The movie was a computer simulation of the paths of Soviet missiles attacking the U.S. By showing thousands of objects streaking across the sky, I hoped to demonstrate the many challenges we uncovered during the Fletcher Study. Unfortunately, the challenge I faced on that day was that the projector the White House had provided for me did not work. James Schlesinger, former secretary of defense, added to the pressure by asking, "How the &#!^$%& can you protect the U.S. if you can't show a *&^%$@ movie?"

Brushing off Schlesinger's concern about my technological prowess, I explained that we realized this ballistic missile defense against an all-out attack was an unsolved problem, but we had proposed a research program to decide whether there was any merit in the concept. I explained that we thought it would take five years and $25 billion to get the factual basis for a decision, rather than simply relying on guesswork. We had no plans to deploy an ICBM force – we just wanted to conduct further research.

Next up was Gershwin, who explained the Soviet Union's commitment to missile defense. His implication was that the Soviets were going ahead with both an offensive and a defensive buildup, no matter what we did. Rankine talked about the proposed requested five-year budget of $18 -$27 billion that was still in development, and how we planned to centrally manage the program in the Department of Defense, with the services acting as the hands-on agents. He also suggested that the services were not very pleased with the DOD calling the shots for this program.

The panel listened to all of this politely; then the serious attacks began. Harold Brown, another former secretary of defense, said he was dead set against defense since it would add uncertainty to the offense. He asserted that the key to deterrence was "the certainty that both societies would be totally and irrevocably destroyed in a nuclear exchange."[43] At that point I thought maybe we could stop the discussion right there, but Scowcroft wanted to pile on. He said forcefully that "the Soviets would just build more offense, and space weapons would be deployed that would destroy our eyes that were critical to all of our military."[44] I realized that the chairman of the president's study had just condemned everything we had presented. Then, James Woolsey, who at the time was a member of Reagan's strategic weapons committee, made light of our efforts and complained about the "mess created by the rhetoric of the speech writers."[45] I began to wonder how the president could have begun this whole exercise without involving the strategic weapons committee, and I thought again, how did I get myself into this mess?

The more they talked, (and they did most of the talking), the more it seemed that what they really wanted were lots of small, survivable, offensive missiles that would preserve the Mutually Assured Destruction doctrine. At the end of the session, Schlesinger became more supportive and said that a new form of deterrence based on defense would be better if it were possible but that there would be a dangerous transition from our present approach. At least he offered some hope that we might be able to make some changes in our military strategy. My feeling at the end of the meeting was that the existing military establishment was very much excluded from the president's initiative, and they might even try to derail it before it even got started. In spite of this lack of support from the president's commission and my immediate reaction that the president's initiative was dead on arrival, it soon became clear there was no turning back.

One week, after the meeting, I got a call from Rankine. He told me that Weinberger's research and development team, managed by Assistant Secretary for Research and Engineering Dick DeLauer, had been asked to convert the Fletcher study into a real request for Congress. I was asked to spend several days in the Pentagon after the Christmas holiday converting the general outlines and the five-year, $25 billion budget into program details and funding schedules. At first I could not believe that we were really going to try to get this budget from Congress with so little support except in the Oval Office, but I was wrong.

The immediate need in the fall of 1983 was to get enough funding together to get the program off the ground in the next year. The venture faced growing opposition, and new money did not exist. The only real way to get a fast start was to scrape off funds from existing programs and relabel them. It seemed to me that Weinberger was playing a poker game, with all the services and the Defense Advanced Research Projects Agency (DARPA) being told to ante up any existing funds related to missile defense. These would be funds already allocated to existing programs, which might never see that money again. In addition, this was a game without defined rules for how to win or lose. As it turned out, nobody really wanted to play the game, but they had no alternative.

Starting with about $1 billion of existing funded programs for FY 84, we were audacious in our thinking and we defined an initial requested budget of $1.99 billion for FY 1985. Doubling any budget in one year was very rare, but something of this size was inconceivable. It was ironic that after the smoke cleared and funds were appropriated by Congress, the funding for FY 85 was $1.4 billion, which was actually less than had been requested by the Pentagon before the president's speech. Nevertheless it was still a substantial increase and a start on a real program, but only if we could build up support for a rapidly increasing budget.

We "guessed" that the five-year "research" program could be finished by 1990, and that would set the stage for a decision about the "what" or if there would be a next step after that. I could not help but wonder which of the two challenges, fusion or strategic defense, was the most likely to be successful, but I had a belief that one was entirely a technical challenge and the other was going to be some unpredictable outcome of mostly global strategic politics and maybe a little technology. We were on the road to more and more investments in offensive weapons unless the technology and some unforeseen developments in strategy and technology could allow us to move in the direction Dyson had suggested in his "Quest for a Concept" speech.

Of all of the barriers to an effective defense, none received more publicity than death rays – weapons that could reach across thousands of kilometers and kill at the speed of light. These futuristic weapons were in reality a very old idea that had been the dream, the obsession and the delusion of thousands of engineers and scientists throughout the world since the turn of the century. I wondered if the dream of "death rays" could turn Reagan's dream of a more stable world into reality.

CHAPTER 14

The SDI program takes off

After spending three months in the summer of 1983 helping to prepare the five-year, $25 billion plan for the Star Wars program, and then in December returning to the Pentagon to help with the actual budget, I had finally gotten back to my work managing the fusion research program at Sandia. I was skeptical that anything would happen with the program. Once again, I was wrong.

In 1984, the Strategic Defense Initiative Organization (SDIO) was officially recognized and a leader was chosen. The leader they chose was Air Force General James Abrahamson, who had been the manager and chief salesman for the F-16 and was the manager of the space shuttle operations. He had a bachelor's degree in engineering from MIT, but his real strengths were his enthusiasm, credibility, boyish charm and communication skills, including the ability to listen with great patience and sincerity to everybody.

In April 1984, Abrahamson was at the Cape preparing for a shuttle launch when he found out his next assignment was to run the SDI program, and, as he often did, he decided enthusiastically that he could get started on his new job while still carrying out his old job. He knew I had chaired the beam weapon part of the Fletcher study and realized that this would be the key to the program, so he contacted me and a few others to give him initial briefings at the Cape.

I had been watching all the publicity about the emerging SDI program and was totally unimpressed by all the contradictory stories of what it really was supposed to do. One characteristic of the spokespeople, who appeared periodically on TV news, was that they all looked uncertain and had the shifty-eyed appearance of used car salesmen who did not really have confidence that the junk car on the back lot would actually run. I was pleased to see that Abrahamson did not have the shifty-eyed uncertainty that I had found common with other SDI representatives. I gave him a quick rundown of the summer study, and he was a fast study and asked good questions. I returned home feeling optimistic about the program and rather disappointed that I wasn't playing a larger role. Perhaps I was caught up in the Star Wars excitement and the lure of higher politics, but my work at Sandia was suddenly starting to lose its appeal. My wife claims I was rapidly becoming bored with the slow pace of fusion research at Sandia, and she was right, as usual. Also, I had gone beyond my five-year limit. Perhaps it was time for a change.

A couple of months later, I put on my dark, pin-striped salesman suit and went to see Abrahamson to see if I could obtain some funding for my lab. I also wanted to see if there was anything I could do to help him become a success in what I thought would be a difficult, if not impossible, job. I found him in a nondescript office in the Pentagon. He greeted me as if we had known each other for ages (I later found out that he had the skill of doing this for everybody), and I asked, "Should I call you general, or General Abrahamson or Jim…or …?" He replied, "Call me Abe." We chatted a bit about the hard technical and political problems, and I noticed his keen blue eyes penetrating my consciousness with his characteristic sincerity and optimism. So I forgot about my sales pitch and said straight out, "Is there anything I can do to help you?" Abe replied, "Gerry, be my deputy director." I was probably suffering from some sort of hypnotic spell, so I said, "Sure, and when do you want me?" His answer was "tomorrow," and mine was "OK" – leaving many small details to be worked out. What about my family? Where would I live? Who would pay me? Perhaps the details weren't actually that small.

The first step I took to resolve those unanswered questions was to call the director of Sandia Labs, George Dacey, and explain what I had just committed to doing. As I recall, his response was something like, "You WHAT? Are you joking?" There were many issues that had to be resolved, such as what would happen to my lab job and who would pay my salary if I went to work with the SDIO. I asked Dacey if Sandia would lend me to the Pentagon to become Abe's deputy for a couple of years, and then I would return to finish up with our $56 million investment in PBFA, and he said, "NO WAY."

He knew that in order to work for the Pentagon I would have to break all ties with Sandia, go work for the government permanently and avoid all conflicts or even appearances of conflicts of interest. His counterproposal was to lend me to the Pentagon as strictly a technical advisor, with no procurement authority. I immediately got back on the phone with Abe. His response was calm, cool, serious and instantaneous. He said, "OK, come as the SDI chief scientist, and by the way, you can also be the acting deputy director until we find somebody else." So I went back to Dacey and he grumbled, but reluctantly agreed. I had resolved some of the details, but the hard part was still facing me: I had to tell my wife.

Much like George Dacey, my wife reacted with, "NO WAY," but after pointing out that we needed to find a place to live, a school for my youngest daughter and an equitable financial arrangement regarding my salary, she was ready to support me. She said, "If you really want to do this, we can work it out."

The clock was ticking, and Abe wanted me on board immediately to start selling the program to supporters, supplicants and detractors. I started almost immediately by traveling to and from Washington. Bob Rankine was the acting deputy to Abe and he had to go to another assignment, so he was anxious for me to get started. I had to wrap up things at Sandia, including putting a new person in charge of the fusion program that I was managing, and I had to find a house, find a school and satisfy my wife that this was really a good idea. Time began to run out with little progress, so on a trip to Washington I set aside three days to get all the details worked out. I arranged a legal and formal method for Sandia to lend me to the Pentagon for exactly two years, I found a school, I found a house to rent and I found a tenant for my existing house in Albuquerque. I made preparations to move my wife, youngest daughter, two dogs and two cats to the Marriott in D.C. while our stuff was being shipped cross-country. My oldest daughter had been accepted for her freshman year at Stanford, so at least that arrangement was made, but now we would be a lot farther from her than we had planned.

CHAPTER 15

My SDI initiation

I BEGAN MY NEW JOB at the end of July with a shocking realization. This was a startup with no bells or whistles. I had come from a very well established, bureaucratic, conservative and orderly place where everything was done according to the rules and regulations. At Sandia, a guard had to touch my badge when I entered the facility. I was always asked when I left if I had any classified material in my briefcase. Certain work could only be discussed in certain rooms behind certain locks. The SDI and Sandia, however, were in two different worlds.

When I arrived at our offices at the Matomic Building at 1717 H St., I recognized the same dilapidated office building where I had presented my first fusion program proposal to a review committee in 1976. Nothing had changed. I am not even sure they had indulged in any janitorial services. There was no air conditioning, no security and very little secretarial support. I walked into my new, cavernous office area, dripping with sweat in the stifling humidity, to find that the lights had been left off to avoid raising the temperature even more. In the gloom, I dimly saw a secretary hunched over a typewriter. I walked up to her and said, "I am Gerry Yonas, the new deputy director and chief scientist," and she sobbed, "Thank God you are here." I did not find this greeting to be reassuring – not in the least.

The building had no security, and my giant office had three doors leading to the pubic corridor. I noticed that each door had a mail slot, and I wondered if one was for confidential, one for secret, and one for top secret mail. Sadly, I was not mistaken.

My first task was to locate the hundreds of proposals that had been sent in seeking funding from the new program. I found an army colonel who said all the proposals were in room 309. That sounded like a promising indication of some discipline and order. We went down the stairs to room 309, which surprisingly was unlocked, and there I saw two dozen safes – all open and brimming with documents, each with its characteristic boldly colored red and white stripes indicating various levels of classification from confidential to secret. Coming from such a rigorous security environment, I almost had a mild seizure but maintained my composure and calmly asked the colonel why the safes were open. His answer was perfectly logical. The safes were sent over in a hurry that day so I could read the documents. By keeping the safes open, I could read them, but of course, I realized they were not secure. Here in one room was the collective technical wisdom of the United States that underpinned the most important new military program in the world. I just stood there for a few minutes and then calmly asked the colonel to rent an 18-wheeler, load up all the safes and drive it continuously around the beltway, occasionally allowing the appropriate people with the necessary clearances to jump onto the moving truck, read the needed documents and then jump off the moving truck again. He looked at me as if I were from Mars and explained to me that such a thing would not be possible. "In that case," I said, "lock every safe right now."

My next task was to meet the two dozen people who had joined the organization. I discovered that most of them were actually working for other organizations and had been loaned to the SDI – just like me. I also learned that they were generally demoralized, confused and upset about the organizational issues, space, hiring slots, budgets, administrative support and lack of direction. They all wanted to know if this was a research program or an early deployment program ... and they were mostly researchers, so I held a staff meeting and tried to put their minds at ease. I declared, "This is a long-term research program to determine if there will ever be a development program." I knew there were lots of advocates of early deployment hiding in the Washington bushes ready to pounce on me.

As July turned into August, I was engaged in reviews of ongoing service activities, talking to organizations already reviewing our work and doing a lot of hand-holding with the staff. Congress was already cutting our budget for FY 85, and we were already making commitments. The Office of Tech Assessment (OTA) was reviewing the program and had already made it clear that we had not taken the obvious countermeasures seriously. I generally agreed with their conclusions, much to the dismay of those in my own organization who wanted me to fight back, and I knew we needed a "credible red team" to out-think the opposition across the river as well as the enemy on the other side of the world.

General Bob Rankine, who was the temporary deputy director, departed at the beginning of September. Bob did not give me much in the way of advice, but he did give me a welcoming present. He gave me his state of Virginia license plate with the label SDI GUY.

General Bob Rankine welcomes Yonas to his new job by giving him his state of Virginia license plate with the label SDI GUY.

Abe was also too busy to give me any information to process, so I was pretty much on my own. One way I could always find out what Abe was thinking was to go to his many presentations, so I went with him to his presentation to the Armed Services Committee. What I learned was that he thought the Soviets had a strong commitment to defense already and that they would agree to reduce their offense if we built a similarly strong defense. Abe's comments indicated that maybe we could also do something to protect our own ICBMs, but we were not well prepared and needed a high level advisory group to help us with all the issues. What Abe wanted was support and well-known big names to help with the marketing. He did not want anyone to review the program or provide technical advice. Whenever I suggested impartial reviewers, he would counter with supporters who could deal with the opposition.

By the end of September, the issues of dealing with the FY 85 budget and planning for the yet-to-be-determined FY 86 budget seemed to take most of our attention.

The SDI budget draws criticism in editorial cartoons such as Berkeley Breathed's Bloom County.

Clearly, our job was managing interactions with the executive branch, the Congress and the media and overseeing the various sub-programs in the hands of the services. We still had a minimum number of staff members, even for that job, and there was no progress in dealing with logistics. But the outside world would not wait, and SDI conferences were being planned by everybody.

I have to admit that my first few days were a bit discouraging, but there was no time to be too cerebral about the whole thing since I was already facing many speaking engagements to hungry contractors and dealing with detractors. The questions were balanced between, "How do I get money?" and, "Why should we spend any money on this?" The only money we had was reallocated up from existing programs in the services, and we needed a new and very real appropriation. We were supposedly managing a five-year, $25 billion program with a few loaned people, no real offices, poor logistics and lousy security. I was beginning to think I had made the most serious mistake of my life, but I was stuck, so it was time to make the best of the situation.

STRATEGIC DEFENSE INITIATIVE ORGANIZATION
WASHINGTON, D.C. 20301-7100
(1717 H Street, N.W.)

December 1984

DIRECTOR'S OFFICE (SDIO/D)			**SYSTEMS (SDIO/SY)**		
LtGen James Abrahamson (Dir)	416	30057/87	Mr. John Gardner	332	30038/67
(Pentagon Office)	3E333	57060	Col Joe Cox	332	30038/65
Dr. Gerold Yonas (DepDir(Actg))	416	30057/91	Ms. Kathy Pratt	332	30038/68
Col Frederick Sawyer (Exec Ofcr)	416	30057/95	Col Tom Fiorino	322	30034/60
Maj Simon Worden (Spec Asst)	416	30057/89	LtCol Gene Kluter	322	30034/63
Capt Rodney Liesveld (Spec Asst)	408	30057/96	Dr. John Hayes	322	30034/62
Mrs. Myrna Rimer (Asst/Manpower)	315	30032/94	Cdr James Offut	350	30033/69
Ms. Judith Lown	416	30057/85	Mr. Guy Barasch	322	30034/64
Ms. Diana Gentry	416	30057/88	Ms. John Takayama	322	30038/77
Ms. Alice Buttke	416	30057/84	Dr. Richard Bleach	362	30038/77
			Ms. Betty Crook	350	30033/72
ADMINISTRATIVE SERVICES (SDIO/AS)			**SENSORS (SDIO/SN)**		
Mr. Glenn Posey	422	30061/82	Dr. Bill Frederick	362	30038/76
CWO Ricardo Payne	422	30061/83	Maj Lanny Larson	362	30038/75
TSgt John Moore	422	30061/79	LtCol Bill Gerber	362	30038
Sp5 Donna Howard	422	30061/83			
			KINETIC ENERGY WEAPONS (SDIO/KE)		
CHIEF SCIENTIST (SDIO/CS)			Dr. David Finkelman	315	30030/92
Dr. Gerold Yonas	416	30057/91			
			SUPPORTING TECHNOLOGIES (SDIO/ST)		
INNOVATIVE SCIENCE & TECHNOLOGY OFFICE (SDIO/IST)			LtCol George Hess	323	31088/20
Dr. James Ionson	315	30028/30			
Ms. Alice Gehl	315	30029/30	**RESOURCE MANAGEMENT (SDIO/RM)**		
Dr. Dwight Duston	315	30029/30	Col Robert Parker	355	30043/56
Dr. Matthew White	315	30029/37	Ms Alice McCarthy	355	30043/54
			Mr. William Bangert	355	30043/59
DIRECTED ENERGY WEAPONS (SDIO/DE)			Mr. Don Koval	355	30043/48
Dr. Louis Marquet	331	30047/25	Maj George Paul	355	30043/55
Ms. Gussie Hopkins	331	30047	Maj Tom Archer	355	30043/55
Col Frederick Holmes	331	30047/23	Ms Gail Gachowski	355	30043/50
LtCol Douglas Kline	331	30047/24			
LtCol Thomas Meyer	331	30047/22			
LtCol Richard Gullickson	331	30047/21			
Mr. Neil Griff	331	30047/34			
Mr. Robert Strunce	331	30047/22			
Ms. Pamela Granzen	331	30047			

PLANNING AND DEVELOPMENT (SDIO/PD)		
Col Joseph Rongeau	329	30053
Ms. Vada Thomas	329	30053/38
LtCol William Wight	329	30053/39
Maj Steve Geary	329	30053/44
Mrs. Mary Peshak	329	30053/41
Mrs. Sally Vaughan	329	30053/42
CONSULTANTS		
Mr. Robert Mercer	315	30030
Ms. Kathleen Troia		
USEFUL NUMBER/INFORMATION		
Prefix 65		
Area Code 202		
Autovon 294 + last 4 digits		
Security 653-0061		

A 1984 list of the employees in the SDI office shows the small size of the operation.

CHAPTER 16

Life at the Pentagon

AFTER A FEW MONTHS, I began to settle into my job and enjoy most of the challenges, but I was still very frustrated with the logistics of the Matomic building. Abe knew I wanted an office in the Pentagon so I could interact more frequently with him, so he pulled some strings to get me moved. I prepared myself to change my environment from a veritable office slum to the epitome of power, the mighty Pentagon. I received the word to show up for work on the other side of the river, with a parking spot at the River entrance and an office on the third floor of the E ring. I headed over one day at 8:00 in the morning. As I stopped at the light before crossing the bridge, I noticed Bill Casey, the head of the CIA, sitting in the back seat of his big black limo, reading his morning notes. I was clearly becoming a powerful Washington commuter.

I easily found a place to park next to a fancy sports car, walked through security and up to the third floor, and I found my new office. I had a fabulous view of the Potomac and the city beyond that. My morning copy of the *Early Bird*, the collection of overnight news articles related to our project, was sitting on my desk and by 8:30 I was in tune with the events of the world and ready to make things happen. The entire time from my garage to my settling into work was less than 30 minutes and was far more pleasant than many other people's morning commutes. Nobody had told me that the perks of being the deputy to a three-star general were those of a two-star general, but I was beginning to learn.

Abe was constantly on the road, so I had the task of going to most of the deputy secretary of defense or secretary of defense staff meetings. The other attendees were the Army, Air Force, and Navy service secretaries reporting directly to Weinberger. There were several others, including General Colin Powell, Weinberger's senior military assistant.

I quickly became accustomed to the routine of heading down the hall to the secretary of defense office with a soldier standing at attention at the door. As I walked through the outer office, I was directed by one of his assistants where to sit, and I was handed a folder of the most important events that might be discussed. The secretary of defense's desk was on the right of his spacious office with what looked sort of like an old-fashioned living room to the left. There was a large, overstuffed chair against the windows overlooking the Potomac, a couch on the left and an assortment of stiff-backed chairs in a semi-circle. As people entered the room, General Powell, Weinberger's military assistant, sat on the couch, the legislative assistant and the press assistants sitting to the left and right of Weinberger and the various assistant secretaries sitting on the chairs. I was never introduced to anyone, and none of the participants interacted with each other. The entire meeting was about getting a message to or from Weinberger, whom everyone called "Cap." The only true interaction seemed to be between Cap and Powell, which reminded me of the interchange between Johnny Carson and Ed McMahon on *The Tonight Show* and often seemed to be carried out strictly for the entertainment of the others. Everyone else was deathly serious. I was clearly treated as a stand-in for Abe, and the only one in the group who made me feel comfortable was Powell, who radiated a sense of self-confidence and warmth. Cap was not particularly talkative. He slumped in his chair and generally appeared to be just as introverted as Powell was extroverted. The regular routine for each meeting was to get the latest on Congress, budget issues and the press. There was usually discussion about some scandal about to hit the Pentagon. Often, the primary subject of these meetings was to discuss strategies for dealing with Congress or getting funding. The discussions were always tactical, immediate and designed to prepare Cap for the events of the day. I seemed to be the only "techie" in the group and I knew enough not to speak up, so mostly I listened and took notes to be able to report back to Abe. As you can see, I did not quite fit in with the rest of the meeting participants. At my first secretary of defense meeting, I noticed that everybody was wearing black, wing-tip shoes and dark suits with red ties. I was wearing a blue blazer with a striped tie and cordovan loafers. I decided that in the interest of national security, I needed to go shopping that afternoon.

CHAPTER 17

Humor can be risky

Although I had a degree in engineering physics, my new job was primarily public relations. I gave a talk or two each week as I represented the program to the public, the scientific community, the Congress and the military. I was constantly in front of one audience or another, fielding reporter questions and traveling across the country and the world as the spokesperson for the SDI. A large part of my role with the SDI was to market and explain the program and to answer questions from the news media. I felt confident about my media skills since I had a lot of experience talking to the press regarding fusion and electron beam research, but I was about to learn that I wasn't as well prepared as I thought.

The SDI public relations director tried to help me. He told me, "Whenever you are talking to the press, don't pay too much attention to what they are asking. Instead, he instructed, "Always go into a meeting with three messages you want to give and, no matter what they ask, always give one of your three messages. You can cycle through them so you aren't repeating yourself, but stick to your three messages." The news media was our best chance for educating the public about our program. We could not pay for a promotional campaign, but we could leverage the news media to tell the story. Any time I had the chance to interact with the news media, I knew I had to take advantage of the opportunity to get my message across.

Early during my time in DC, I was making a presentation when a young *Los Angeles Times* reporter named Robert Scheer approached me. Scheer was well-known as an anti-Pentagon, fairly liberal reporter, but he seemed very nice and, after my talk, he offered me a ride to the airport. He had a Porsche 911 and as we were driving down the freeway he asked me some questions. What I didn't know was that he had his tape recorder rolling. He kept asking me all sorts of questions about the program, including, "What are you not going to do?" After all, Scheer noted, "You can't do everything." He kept pestering me, so I joked, "We're not going to use a crossbow, because you can't pull the strings tight enough when the missiles are coming." The next day that quote about the crossbow appeared in the *LA Times*.

During the drive, Scheer also asked me how we would be able to shoot lasers into space to bounce lasers off mirrors when the weather was cloudy. I replied, "Of course you can't shoot through clouds – that won't work." I should have said we'd go to another site, but instead I just blurted out that it wouldn't work. The next day the headline in the *LA Times* was "Cloudy Day ruins SDI."[46] During my morning meeting with the secretary of defense, Cap just glowered at me.

To counter the bad press, the secretary asked Abe to do an experiment and shoot a laser into space and bounce the beam off a mirror. Abe arranged this experiment, and it worked the second time we tried it. Then the secretary explained to the news media that it could be done. There was just one catch: we shot the beam off Mount Haleakala in Hawaii, and the mountaintop was above the clouds. If there are no clouds, you can shoot a laser through space. Fortunately, the news media missed that detail.

The press was very involved in what we were doing. Reporters wanted to know if the program would work, how it would work, how much it would cost and when it would be ready. They asked me those same questions over and over again. I knew they had a job to do. My job was to keep them happy without actually answering their questions.

One of the highlights of my media experience was when I opened the newspaper one morning to discover that I was one of the characters in Berkeley Breathed's *Bloom County* comic strip. Technically, the evil SDI chief scientist (who was also a penguin) was the comic strip character.[47] I found this oddly flattering. I still have a framed print of one of the *Bloom County* cartoons in my office today. No matter what we did, SDI had become the focus of late-night comedians and the brunt of their jokes. I was only 45 years old, and I had always used a lot of humor. I soon discovered that when I made a joke, it would backfire – but I often realized I shouldn't have been joking after it was already too late.

SDI becomes the focus of late night comedians and editorial cartoons, such as this Bloom County strip about the evil SDI chief scientist.

Naturally, it wasn't very long before the folks at the Pentagon decided I needed some training. Actually, they didn't tell me I needed training. They told me we were going to have a special meeting with the media in the pressroom in the Pentagon. I didn't know that they had prepared – digging up secrets from the staff that they could throw at me during the interview. In actuality, they were not reporters at all – they were all from the Air Force. During the interview, the lights went down and the reporters said there was a technical problem – so they could trick me into saying the wrong thing while we were chatting during the wait time. At another point, as I came out of the bathroom, a reporter jumped out from behind a potted plant and started firing questions at me. They tried every possible trick. Afterward they admitted who they were and debriefed me. I certainly became aware that dealing with the news media meant you had to be on your toes.

Speaking of being on your toes, I also learned the benefit of sitting up straight. One time I was on *Good Morning America with David Hartman*. During this show, I was debating Richard Garwin. I sat up straight in my chair, and Garwin was slumping. It looked like I was overpowering him and, needless to say, I won the debate.

The challenge of dealing with the news media during that time was that they wanted to hear we had something ready to go, but we really didn't have any answers. I would tell reporters, "We can't make cost estimates until we do research. It will take five years until we know if we have anything worth developing. Until then, it's a research project." My job was to work the line between being an outright liar and telling the truth. I had to be evasive without seeming to be evasive ... to tell the story without telling the wrong story. It was a difficult balancing act.

The only thing more challenging than talking to the news media was talking to Congress. The public relations experts told me when you testify in front of Congress, you must never lie to the Congress – but don't tell them the truth. I gained firsthand experience with this delicate dance when we offered European countries incentives to support the SDI program. I was testifying before a congressional banking committee, and one congressional representative asked me if we were "buying off European allies?" I avoided a problem by delving into complex technical information. I became quite fond of this technique of going into long, detailed technical discussions when I did not want to answer a question. It seems that people would lose interest if I talked technical gibberish so I could dodge difficult questions. I might have looked foolish, but I didn't lie ... and I didn't tell the truth. After the incident with the banking committee, I was concerned that I had lied to the Congress, so the next day I went to the Pentagon's counsel and told him about my concerns. He directed me to the complete transcript of the hearing. After I read it, I went back to the attorney to report that what I said during the hearing was mostly incomprehensible. He replied that there was no law against being stupid or spouting nonsense, and sent me on my way.

As you can see, we had our hands full. The program was wallowing in controversy from the moment Reagan delivered the infamous Star Wars speech. There were so many different opinions about what we should do, by, when, for how much. There was so much at stake. While this environment could have created an atmosphere of deception, instead the tone of the program was one of sincerity, integrity, total dedication and thoughtfulness. I always gave Abe credit for providing the leadership that set the standards of the program. His approach was open-minded to every idea – almost to a fault – and his patience was unreal. He became deeply involved in every discussion and proposal, and he often exhausted the presenters with his probing questions, but he was always optimistic rather than critical. In fact, the presenters always went away even more optimistic about their ideas than when they arrived.

Abe really believed in technological miracles, and he was able to convince others they could achieve more they had expected. I was more of the skeptical observer and would often send him memos with my concerns. He would always respond with his handwritten response in the margins and end his comments with his characteristic signature and a happy face. Can you imagine any other three-star general signing formal technical memos with a happy face? That happy face truly reflected Abe's character and the way his positive approach to leadership set the tone.

Abe never resorted to sarcasm or teasing, so at first he found my sense of humor unsettling. Slowly though, as he began to understand my personality, he developed the theory of "The Yonas Hook" to describe times when I would tease or trick him and he would fall for my ruse like a fish with a hook in its mouth. Abe would occasionally walk into my office when I was meeting with someone and warn that person to watch out for the hook, demonstrating the effect by inserting his forefinger into the side of his mouth.

Of course, I was not entirely blameless. I recall one day in particular when Abe was out of town and I attended Secretary of Defense Weinberger's staff meeting in Abe's place. As I went through the notes prepared for the staff meeting, I read a brief mentioning that my assistant, Major Rod Liesveld, had briefed the U.S. embassy in Paris regarding the SDI. That was the extent of the note, but this energized my bad behavior, and I could not resist a small joke, especially since I had just the right kind of relationship with my assistant.

That afternoon I called Liesveld into the office and scolded him, pointing my finger and telling him I had heard at the staff meeting that he had messed up everything in Europe. "I read that you went to Paris and met with the ambassador and his staff, and you insulted them so badly that you have caused an uproar. The entire alliance is endangered! The Secretary of Defense is incensed! SDI is in danger because of you!" Knowing me all too well, Liesveld just smiled, obviously enjoying the joke, flipped me the finger and walked out. Neither of us noticed that Colonel Frank Sterling, Abrahamson's new executive assistant, had overheard the entire phony incident. The next day, when Abe returned, Colonel Sterling asked him what where they going to do about the Paris situation. "What situation is that?" asked Abe. Sterling went on to describe what he had heard.

"Oh my God!" shouted Abe. "I must advise the secretary immediately! I must apologize!" He ran off down the hall toward Weinberger's office. Overhearing the commotion, I asked Sterling what was happening. Sterling replied, "General Abrahamson was on his way to see the Secretary of Defense regarding the Paris incident." I swallowed hard, turned white and decided it would be best for me not to be found, so I headed for the cafeteria, thinking my SDI days were over. As Abe went toward Weinberger's office, he almost ran over Colin Powell, Weinberger's executive assistant, who saw the worry on Abe's face and asked him what was going on. A hyperventilating Abe went on to describe what he had heard. Powell, who would have known if such an incident had indeed occurred, smiled and said, "Can't you see someone is just pulling your leg, Abe!?" As Powell walked off, Abe remained there pondering for a moment and then realized who was to blame for the situation. He ran down the hall and stormed through the doors of my office, past my secretary, bellowing, "Yonas... Yonas... Yonaaaas!" He had been taken by the Yonas hook.

On another occasion, I was giving a briefing at the National Security Agency. As I entered, the security guards advised me that I could not bring anything in or take anything out, so I slyly slipped a fancy paper doily stamped with the NSA logo into my pocket in protest. A few weeks later, I met Abe for a drink (cranberry juice for Abe) after he gave a dinner speech to discuss the audience reaction. As I fished into my pocket for something to write on, I felt the doily with NSA's logo. When Abe was not looking, I placed the doily under his juice glass, and when he noticed, I feigned surprise. "Abe, there seems to be something under your glass." He picked it up, surprised by the NSA emblem. "The National Security Agency – what can this be doing here?" I said, naively, and then gasped in feigned horror. "Someone is sending you a message, Abe. Someone must be trying to tell you something!" Abe looked around for a second, trying to see if anyone was watching, but suddenly he stopped and looked back at me, pointing a finger. "You, you, you and that hook!"

Abe rarely tried to get revenge for my trickery, but once, when we were visiting the State Department, he noticed a homeless man sleeping on a grate near the sidewalk. Without missing a beat, Abe pointed him out to me and said, "You see, Yonas, even the State Department has a chief scientist." It had taken awhile, but Abe had finally had given me one back.

Lt. Gen. James Abrahamson reacts with amusement as Yonas presents him with an award.

CHAPTER 18

Dealing with the defense community

IT APPEARED TO ME THAT the American defense community knew nothing about the SDI program. In fact, it seemed that most of those in the defense industry just wanted to protect their budgets. My first opportunity to meet with leaders in the defense community and share ideas was the "Mitre C3 National Security Symposium" in October 1984. I knew there would be a great deal of emphasis on arms control issues at the conference, and I was particularly interested in hearing how Abe would handle these complex issues in front of a truly well-informed audience. Abe did not hide his strong beliefs. He claimed, "Tech miracles can happen. We can do anything we want to do and, by the first decade of the next century, we will have conquered disease, misunderstandings and lack of communication." Abe said our job required the ability to kill boosters early in their flight and that we had the tools to do so. "The free electron laser is real, the chemical laser is real, rail guns are real, neutral particle beams are real. We even have a neutral particle beam weapon operating today ... but a small one," Abe said. He went on to comment that, with a credible defense, we could reduce the value of offense and thus negotiate much lower levels. He made it clear that the goal of the "long-term program was arms control and enhanced deterrence." In the questions that followed, one person asked how we would respond to budget cuts and Abe responded, "With a cut, we will do it faster and cheaper ... go faster and faster and faster."[48] This reminded me of the fighter pilot advice he had given me once: "When all of the warning lights in the cockpit are flashing red, pull back on the stick, climb and go faster." His motto was, "If you wait to make a decision, it is too late." Personally, I would have bailed out of that plane, but since I had no parachute in my metaphorical plane, I decided to go along for the ride and have some fun.

Abe was followed at the conference by Ash Carter, a 30-year-old physicist and Rhodes scholar. I was impressed by his technical analysis and policy perspectives, and I expected he would provide a clear view of the many SDI issues. Carter noted that space would be a very dangerous place and would need enforceable agreements to protect warning satellites and some undisclosed way to protect satellites that would be used for targeting. He advocated "patience, continuity of thought and coherence,"[49] and I thought all were in short supply.

Professor Steve Meyer, an MIT expert on the Soviet defense establishment, followed with a focus on the Soviet view of SDI. He claimed they believed we knew what we were doing, but that they would be first to deploy a laser weapon in space and would also respond with an offense buildup. He made it seem that we were on the path to eliminate any opportunities for arms control and that the future would be very confrontational.[50] It seemed to me that we were going down the path toward war in space. Star Wars was a good description of the future.

Other speakers discussed issues of space control, commercial space and space "rules of the road" without any explanation of relevant technology. Professor Al Carnesale then warned all of us to be cautious, slow down, not make promises and try to protect the ABM treaty. He believed that space survivability, cost and stability would doom SDI, and he suggested that if we wanted to go ahead with a deployment, we should keep the good stuff but share anything that did not work with the Soviets.[51]

We were planning to spend $3.8 billion in FY 86, and I suspected if Congress thought we were going to share our results with the Soviets, we would get nothing. In the best of situations, the pressures on the budget were growing and the credibility of the program was in serious trouble.

In November 1984, I met with JASON, a group of notable academic scientists, mostly old white male physicists, who would get together in La Jolla, California, during the summer months, which is where the group got its name from an acronym for July, August, September, October and November. The group's goal was to evaluate programs and technology for the defense department. They also conducted summary meetings in the spring and winter to share the results of their studies with each other and the defense community. I had attended many of these JASON review meetings over the years and was particularly fond of the cocktail party on the Friday night before the Saturday meetings. This was where you could find out what the academic community really thought about the new science and technology challenges in the Pentagon.

There was no question – the JASON members were highly skeptical of the SDI technology claims and were convinced that the outcome would not be favorable to arms control, which was their primary goal. One of the key leaders on the subject of science and technology/arms control was Stanford physicist Sidney Drell. He made it clear that the management and understanding of defense problems was "worse than ever." He claimed there was a tendency to deploy systems before even the basic physics was known and that decisions were being made on the basis of "self-delusion and hype." He recommended that we work on short wavelength lasers and not chemical lasers or particle beams.[52] This incidentally was the same conclusion we drew in the Fletcher study in the summer of 1983. He also said the Soviets would invest in anti-satellite weapons and penetration aids to defeat our defense. I think he captured the sense of the JASON community that we needed advice from a standing impartial academic panel. I could not argue with most of their opinions, but I did think that maybe it was time to introduce some new young, diverse people into the JASONs. It also seemed to me that the outcome of our efforts would to a large degree depend on the Soviet reaction and perceptions rather than on reality, but little was known about Soviet psychology. I really thought we were engaged in a mind game rather than a technology game, but all the experts preferred physics rather than human behavior. Physics is so much less complex.

On Dec. 4, 1984, the "Roundtable Discussion Focus on SDI" took place at the Air Force Association Headquarters just across the Potomac River from the Pentagon. I assumed this would be a love-in with audience support from the aerospace industry and the military and no physicists. I had no information other than to show up and talk about the SDI, and though I was surprised that nobody had bothered to prepare any of us for the discussions, I had become used to surprises. The other presenters were Pentagon professionals, one congressman and a Pentagon correspondent. As it turned out, I found that none of the others really knew much about the SDI beyond what was in the president's speech the year before.

The meeting moderator made it clear at the outset that funding was in doubt, the program was poorly understood and the purpose of the meeting was to convene a jury to judge the program. He emphasized that the objective of the SDI was the "creation of a leak-proof defense ... and there would be no further need for continuation of the strategic force modernization program." He went on to say we should not worry about SDI competing for funds since Congress had already cut $377 million, leaving only $1.4 billion for FY 85.

Since we were already planning on $3.8 billion for the following year, I knew that we would have to drastically cut back some programs.

The first speaker, T.K. Jones, undersecretary of defense for research and engineering, strategic and theatre nuclear forces, started us off with some bizarre comments, saying, "Defense is the only remaining alternative to nuclear winter, and SDI offers a plausible response to the morality and the nuclear winter issues."[53] This was the first and last time I heard any of the experts mention morality and nuclear winter and, frankly, I was confused.

The next speaker, Frank Gaffney, who I knew from his work in the policy office in the Pentagon, argued that the "vast majority of Americans continue, quite sensibly, to fall back on the survival instinct that says we want defense."[54] Gaffney said defense was "an essential building block to a more stable, more secure strategic and arms control posture." I could not find anything in his description that was different from the public statements I had been making. At least the policy shop and our office were synchronized.

Then came a series of talks indicating that the Pentagon establishment was very far away from providing any support for the SDI. Don Latham, the Pentagon leader for C3I, Command Control Communications and Intelligence, raised the issue of a system on automatic pilot and made it clear there would always be a human in the loop. I wondered if that human were the president, how could he always be prepared to make the launch decision in a matter of minutes?

Brigadier General Don Kutyna, the Air Force director for space systems, was up next and his task was to explain how the vital space assets were to survive in the face of space mines, spoofing the commands to the satellites, terrorists destroying ground facilities, blinding space nuclear explosions, laser threats, electromagnetic pulses, jamming the communication link and the need to rapidly replace destroyed capabilities. He offered no answers and decried the dependence on the four space shuttles for deployment of our military space hardware. None of us knew how right he was, but just one year later, the space shuttle Challenger blew up shortly after launch, and the fleet was grounded for three years.

When it was my turn to speak, I had no answers ready for the doubters and detractors. I was well-known as an expert of sorts in directed energy weapons and I made it clear I had my doubts about lasers, but I advocated kinetic energy weapons, which I called smart rocks. I emphasized the need to kill the boosters before they could deploy their payload, and I said we needed to prepare for the near term Soviet response in 10 years, the midterm in 20 years, "but also figure out what they would have in the long term future." The audience appeared rather confused and some people even seemed angry, but then U.S. Rep. Ken Kramer took the microphone. He had obviously thought about all of this complexity and appeared ready to provide the answers to the anxious audience. Unfortunately, he proceeded to spout nonsense and leave everyone more confused.

Kramer claimed that "the Soviets have under development a heavy launch vehicle that is anticipated can take to low orbit a modular space station of very ambitious proportions that might house as many as a hundred cosmonauts within 10 years ... and they have spent about three to five times as much on their directed energy research and development program as we have." I could not imagine why anybody in their right mind would want soldiers in space. I guessed he had sat in on some classified briefings that I had missed. Kramer went on to say, "We have taken a lot of important steps in the last couple of years to rethink that vision – that dream – a potential reality in the not too distant future."[55] Since the program had only been up and running for a few months, I wondered how I had missed all of that?

Kramer said that SDI "may be the single largest undertaking ever done by mankind." He claimed the Fletcher study estimated a cost "upward of a trillion dollars-plus" and said, "Money is simply money ... given the threat of nuclear weapons." At this point I felt that I had lost track and that I was certainly inadequate to be the chief scientist for mankind's largest undertaking. Like the rest of the audience members, I was becoming very confused.

Next came Al Pierce, the NBC reporter and experienced Pentagon insider, to make sense out of all of this. Pierce stated that the president was "winging it ... with very little consultation with a lot of the senior civilian and military experts."[56] I knew that was indeed the case. Pierce said we should decide what we were going to defend and what we were going to defend against and argued that we should figure out how to protect our ICBM forces." I knew this was certainly wrong since we had rejected any thought of a defense of our ICBM's as contrary to our goal of shifting from offense to defense. Clearly, no one truly understood what we were doing

In the discussion that followed, the moderator told the congressman that "claiming that defenses would replace offensive forces is both dangerous and wrong." Kramer disagreed but did not argue his case. Jones then made it clear that SDI was research, "and it is a good 10 years before you get to the leading edge of SDI" and we need to keep on spending on "our security." These comments suggested we would need increasing defense budgets for several years to pay for both offense and defense buildups while paying for a lot of research. All of this seemed unlikely in the light of the congressional pressure to cut defense budgets.

The only redeeming aspect of this meeting was that sitting in the front row was none other than Jimmy Doolittle, who I had always admired as one of our greatest war heroes, so I went down and got his autograph in my notebook. In my two years in Washington, I met with many famous people, but he was the only one I ever asked for an autograph. So much for conferences. I concluded that the "real" defense community had little interest in and less understanding of the SDI. I was sure they were primarily interested in protecting their budgets from the grasp of this greedy program at a time of increasing pressure on the defense budget.

CHAPTER 19

Facing our adversaries

THE SDI FACED A NUMBER of serious adversaries, and one of my most challenging duties was figuring out which battles to fight. I quickly learned that one of the greatest threats was not to be found in the Soviet Union, but just across the Potomac. It was Senator William Proxmire, who loved to embarrass us for wasting money – pointing out purchases of $100 hammers and $10 bolts. One day I learned that we were struggling to decide how to handle a most shocking expenditure: $7,000 coffeemakers for aircraft. Yes, we could argue with Proxmire that the coffee pot could make 90 cups of coffee in two minutes, survive a crash and make coffee for any survivors of the crash as well as for the first responders, but we feared what would happen if Proxmire discovered the coffee pot's cost. One morning, John Lehman, the Secretary of the Navy, came in sheepishly, followed by an assistant holding a toilet seat. Actually, it was really more of a toilet seat cover that cost $600, and Proxmire was planning a news conference that day. The room buzzed with consternation and confusion. Weinberger just sat there silent and slumped in his overstuffed chair, not looking at the rest of us sitting stiffly in our rigid uncomfortable chairs. After what seemed like an infinite amount of complaining and hand-wringing, Weinberger looked up and announced, "Gentlemen, we will sit on this story."[57]

Not every crisis could be resolved with humor. Another of our adversaries was confusion. The president had made it clear that he wanted to abolish nuclear weapons, and this raised a real contradiction with the X-ray laser program advocated by Edward Teller. Reagan had a strong relationship with Teller and a visceral dislike for nuclear weapons, so Cap would frequently describe the X-ray laser as "nuclear-powered" or "using a nuclear generator" but never mentioned the fact that the proposed X-ray laser was powered by a very large nuclear explosion in space. This struck me as intentionally misleading people, and I felt it made the secretary look uninformed. I became increasingly frustrated and one day took my complaint about the secretary's confusion to Powell. I asked Powell if he could straighten out the story. I also told him that the Teller story about "popping up" the nuclear weapon into space from a submarine just when it was needed would make no sense in the face of a Soviet countermeasure using a fast-burn launching missile. I had concluded that there would have to be many nuclear weapons permanently deployed in space, which would never be acceptable to either side. Powell's reaction was straightforward, brief and dismissive, as if to say, "The secretary knows all of that, and he is not worried about the details." This ability to think more about the political implications than the facts was something I never became accustomed to. I soon learned to be careful about who I irritated by bringing up facts.

Another challenge was learning to recognize the enemies. One day we had one of our very infrequent meetings on overall strategy in an out-of-the way, small, carefully controlled conference room. Two of the principle strategic thinkers at the meeting were Frank Gaffney, who worked for Richard Perle, and Paul Nitze. Perle was a brilliant theorist and a totally anti-communist senior Pentagon policy analyst, and Gaffney was his point man for SDI issues. Nitze was the highly respected senior strategic advisor to Secretary of State George Schultz. The purpose of the meeting was for Nitze to explain what became known as the famous Nitze SDI criterion: any defense deployment would have to be cost-effective compared to the likely Soviet offensive response. I had already stated that obvious observation in various speeches and articles, so when Nitze said it, I immediately supported him. Gaffney was visibly angry about my comments and after the meeting, I asked him to explain. He said, "If you ever support the enemy again, your days in the Pentagon will be few." I responded that the enemy could not have been in that closed and secret meeting, and he shouted, "You know the enemy is the State Department," and then turned and stalked away.

I learned not to ruffle Gaffney's feathers too much, but occasionally I could not help myself. At meetings we would go around the conference table explaining our roles and, when I followed Gaffney's grand explanation of his important role in the Pentagon, I explained that my job was to help Gaffney with the technical facts. Naturally that angered him, so I tried hard to be respectful – most of the time. I also did not want to get on the wrong side of Perle, the so-called "Prince of Darkness." As time went on, I grew to respect Perle's ability to manage complex situations with intellect and style. I recall the time Abe suddenly left me in charge after learning that his wife's plane was missing and had probably crashed in the Sierra Mountains. Abe had left me with the task of testifying the next day in front of the banking committee, which was after us for making financial deals with our allies without any Congressional approval. Fortunately, I was sitting next to Richard Perle, who jumped in when I was accused of bribing the allies for political support. I started to talk about technical challenges, and the chairman yelled at me to answer the question and "deny for the record that you have not been buying support from our allies." As I mumbled something incoherent, Perle jumped in and attacked the committee for some sort of banking indiscretion, which allowed me to become mute and invisible instead of being verbally eviscerated.

From then on I was convinced that Perle could deal with any verbal attack and respond with a vigorous defense and emerge victorious. That is, until he answered a call during a meeting with a group of us late one day in his office. During the call, he was totally apologetic, asking for forgiveness.

After he finished the call, he explained to our group that the caller was his 8-year-old son, and he had forgotten he had to be somewhere else at that moment.

I had a lot to learn about the realities of Pentagon politics, military culture and dealing with Congress, but with regard to one subject, I was totally ignorant. Because I had never served in the military, I was never aware of the privileges of rank. One day, a neatly dressed and very professional-looking Air Force lieutenant colonel showed up in my office and introduced herself as Jean Ostreich, the executive officer for the deputy director, who was me in my acting role until we found a replacement. She informed me that she would provide needed support functions. I was dumbfounded, because I already had a secretary and a very competent military assistant. I had no idea what to do with an executive officer, so I went to Abe's executive officer, Colonel Frank Sterling, and asked him for instructions. He said, "Do as you are told, and she will take care of everything." I decided to try it. The next day she told me about the arrangements she was making for "our trip to Hawaii." I could not imagine going to Hawaii with a woman I barely knew, and I was concerned about how to explain all of this to Jane, my wife. Before I could talk to Jane, Jean called her to let her know that she was making arrangements for our activities during the day and wanted to know what I liked to do at night. That night I was faced with the marital equivalent of nuclear fallout, so the next day I decided I needed to have a talk with Jean. Unfortunately, my use of proper military protocol language was inadequate, and she just stared at me like some kind of uneducated civilian.

Eventually Jean and I learned to get along and I became accustomed to having her there to make arrangements, prepare notes, carry my briefcase and take care of everything else. One day she asked me why I did not fly military air, which for a two-star meant I had access to a Learjet when it was available. Because I often attended meetings at the Air Force research lab in Albuquerque, which had no direct service, I decided to try it. Wow, was that nice!

Once I was meeting with Army and Navy brass and the meeting was running late and they had to catch their commercial flight, so I asked Jean if we could offer them a ride. Talk about a "one up" in the service rivalry. On my next trip, I flew in the Learjet to save time and, when we landed, I saw a general standing at attention near the plane door. I asked Jean who was the big shot on the plane. She replied, "You." It did not take long for me to become totally dependent on Jean for all arrangements, and when I left the Pentagon to return to a normal existence, I was practically helpless. It took months to relearn normal human coping skills.

CHAPTER 20

An alien accusation

As my work on the SDI program continued, I found we faced an unexpected and very strange adversary in the form of a well-known newspaper editor and publisher. In 1985, Duke Tully, from the *Arizona Republic* and *Phoenix Gazette,* was one of the most powerful political 'kingmakers" in the state of Arizona. His newspapers had a combined circulation of 400,000. One aspect of his fame was that he claimed to be a decorated Air Force pilot who had crash-landed his P51 in Korea. His saga involved a year in a hospital where he had his smashed front teeth replaced with teeth of steel. His military record included 100 missions over Vietnam. He had the Purple Heart and the Distinguished Flying Cross and often wore his lieutenant colonel uniform to banquets and fund-raising events.

In actuality, Tully had only been in the civil air patrol and had never served in the military, but he had many medals, phony plaques and awards that he displayed in his home and his office. His good friend, Dick Rose, a freelance aviation reporter with the *Arizona Republic,* said, "Tully talked the real military language, his demeanor was military, and his stories were military accurate." Tully was described by the *Chicago Tribune* as "equal parts cowboy, commando, swashbuckler and elegant tycoon, who was a George Patton who drove a Corvette, a Randolph Hearst who flew an F-16, a John Wayne in aviator glasses and Air Force dress blues." Tully had groomed John McCain for public office, and McCain had asked Tully to be the godfather of one of his children. Tully was also close to other figures in Arizona such as Senator Dennis DiConcini and the gangland boss Joe Bonano, who had retired in Tucson, Arizona.

Although Tully had lived a lie successfully for many years, in 1985, the pressure was getting to him. He began to drink and had alienated his second wife, Pat, who filed for divorce. In October 1985, he was starting to crack up, and his aide found him in his office "stepping on his plaques and throwing them into a trash can."[58]

Dick Rose, Tully's friend due to their mutual interest in aircraft, was a freelance reporter who was not particularly educated in highly technical subjects. In the early '80s, Rose became fascinated by the idea of magnetohydrodynamic (MHD) power generation and propulsion. The idea seemed to be a route to a solution to energy problems and, when extrapolated to extreme conditions, Rose thought it might lead to compact power supplies and even beam weapons. Undoubtedly he was influenced by the speculation of scientists and engineers that, by combining many sequential miracles, vehicles could be created that would defy gravity and fly at hypersonic speeds. He became so fascinated with the possibilities that he convinced himself that the applications of MHD would "alter the course civilization on planet earth ... would rival the discovery of fire, the wheel, gunpowder and the atomic bomb ... and were already known in high government circles."[59] Rose expected to reveal all of this in a story he had written, and he had shared this story with Duke Tully in early 1985.

At this point in his rapidly disintegrating life, Tully needed a miracle to save himself, and when he heard the MHD story from Rose, he took it to his friend, Senator DiConcini, who was on the Senate select committee on intelligence. DiConcini had access to some of the nation's most tightly controlled secrets, and he certainly could confirm or deny this story. After listening to Rose, DiConcini also became fascinated with the possibility of using MHD as power for space propulsion and directed energy weapons, since such technologies were a very hot topic in Washington following the initiation of the SDI program. Stranger than fiction, Rose believed that "recirculating a liquid metal through the MHD duct would result in an extraordinary amount of electricity," and he convinced Tully and then DiConcini that the U.S. already had been operating the "lightening space craft that used the liquid metal in the MHD duct to cancel the effect of gravity."[60] Because this sounded like some sort of UFO, Rose argued that the top-secret program had been moved to Kwajalein in the South Pacific to protect it from premature disclosure.

DiConcini wanted to get at the bottom of all this and suspected that the SDI program would have the answers, so he invited me, the chief scientist who was also, according to Rose, an expert in MHD, to visit his office to answer questions about what he described as a secret alien spacecraft on a Pacific Island. I knew a bit about MHD since I had studied the subject in my undergrad thesis at Cornell. I also did my Ph.D. thesis at Caltech, which included experimental research in 1962 to 1967 on a liquid metal MHD duct at the Jet Propulsion Lab. I have to admit, however, I had not thought about the subject for even one millisecond once I left JPL in 1967. Given Rose's obsession with MHD, who could be better to reveal the real secrets of MHD propulsion than Gerold Yonas, who had become the leader of the directed energy weapon study for the Fletcher Report in 1983 and then the chief scientist and deputy director of the SDI in 1984.

DiConcini called me into a meeting, which turned into a disaster of confusion and poor communication because the two of us were on two different intellectual planets and I just could not get serious about the discussion. The senator never mentioned MHD but asked me directly if there were an alien spacecraft on an SDI-managed facility on a South Pacific Island. I respectfully said no. "So why won't you let the reporter Dick Rose from my state's most prestigious newspaper go there and just take a look around?" DiConcini asked. Because the Pentagon had refused to allow the reporter to visit, DiConcini was certain we were covering up important information. I denied there were any alien spacecraft anywhere in the U.S., and certainly not at any SDI facility. I then returned to the Pentagon convinced that my life had been transformed into a grand joke.

Following this meeting with DiConcini, I wrote a silly memo to document what had happened – paying particular attention to the parts about aliens and islands and embellishing it a bit, writing, "Garn the horrible, the only remnant of a lost tribe of space pirates, was launched into orbit in search of funding. Crazed by a lack of media attention, he threatened to beam up Danny Graham, Senator Wallop and Edward Teller if we did not satisfy his demands." Assuming Abe would enjoy the humor and figuring that the entire episode was just another crazy encounter in the SDI world of extremes and silliness, I promptly put it out of my mind.

But this was just the beginning of the Duke Tully story. DiConcini, totally miffed by my lack of serious attention to the subject, then summoned Abrahamson to come back with a better answer to his questions. One day Abe, without any warning, grabbed me and Colonel Sterling to meet immediately with DiConcini to provide him with a more complete answer. Fortunately, the silly memo that I had written after the first visit never got to Abe, and this time, after we listened to the senator's story without laughing, Abe told DiConcini, "We will get to the bottom of this."

I was stunned by Abe's reaction, but he was more adept at dealing with the politics than I. Unfortunately, this left the door open for a follow-up in a formal meeting with DiConcini, Duke Tully and the reporter Dick Rose on March 19, 1985. As it turned out, the meeting in the Pentagon turned out to be a trial, and the person on trial was none other than me, since I had done pioneering work on MHD propulsion but did not have any hint that there was any connection with alien spacecraft. Before the trial was over, I was accused of being in cahoots with aliens and joining the SDI because of my work for the aliens.

I was amazed that a senator took all of this seriously, and I could not help myself from laughing, so I abruptly told the visitors, "I have to go to Texas," and fled from the room – leaving Abe and the Air Force colonels to handle the questions. Rose told Abe the story he was going to publish would connect Tully, DiConcini, Abe and the unwitting alien co-conspirator, Dr. Gerold Yonas, in the story of the century.

Rose sent his completed story to the Pentagon for review in the weeks following the meeting. I was asked to review it for classification. The article reported the following dialogue that hinged on my answer to the following question: "At what stage of development is the rotary magnetoplasmadynamic propulsion system?" My answer, as Rose reported it and as I remember it, was, "I don't think I have ever heard of anyone working with such a concept." Rose then wrote, "It was Gerold Yonas who had begun work with this futuristic MHD variation during his graduate studies … one of the forerunners in the technology … one reason he was selected for the position of deputy director of

SDI." Rose also wrote that an MHD-propelled spacecraft had "been recovered virtually intact from a crash site in the New Mexico desert ... remains of the alien beings were recovered as well."

In the article, Rose described asking Abe, "Is there any truth to these stories of a ship, not from this earth, being examined by the Air Force?" According to Rose, Abe said, "I have no personal knowledge of a project involving an alien spacecraft ... it may not be an intact ship but some bits and pieces." Rose then asked whether research projects centering on an alien spacecraft provided technical information that allowed American scientists to move their knowledge of space sciences ahead a hundred years or more. He quoted Abe as saying, "Our intelligence community is working on a project of communicating with outer space intelligence." Rose also claimed that the Russians were close on the trail of this work and in 1984 had flown the first prototype of their spacecraft.[61]

After reading the draft of the article, I decided that there was no law against publishing silliness. I believed I had more important issues to deal with, so I gave permission for him to publish it. I also thought it would be fun to be faced with answering questions about working with the aliens.

But this really entertaining story never appeared, and, when I asked the Pentagon officials about the publication, they said the reporter had been blown up in a huge car explosion and "was vaporized," which seemed no sillier than any of the rest of the story. As a result, the word got around in the Pentagon not to mess with me. By the end of 1985, Tully had skipped town after his exposure as a fraud, and Rose was never heard from again. The story still exists in my files, and I sometimes wonder if the aliens might want to vaporize the only copy of the story – or even vaporize me.

CHAPTER 21

Edward Teller and developments in space wars technology

IN THE EARLY '80S, THERE were three prominent weapon concepts for ballistic missile defense. Danny Graham's concept involved thousands of small missiles deployed in space and was based on an old idea from ARPA that had originally been called Ballistic Missile Boost Intercept or BAMBI. The space-based chemical laser funded by DARPA was touted by the aerospace companies and championed by Senator Wallop. This concept involved dozens of 100-ton laser platforms fueled by chemical reactions. Finally, there was Edward Teller's X-ray lasers driven by nuclear weapons deployed in space. I could not begin to imagine how we would solve the practical problems of getting all the hundreds of tons into space for the chemical lasers and the small missiles, although I could imagine solving the purely technical issues of deploying dozens of small nuclear bombs in space. Of course that possibility was not only a violation of the 1967 outer space treaty but seemed to be something the Soviets would never accept and something that could lead to a real war in space. I also felt that although the ideas were theoretically elegant, the practical applications could not be achieved. We also had to overcome the public opposition to the idea of "Nukes in Space." I remembered that Teller had two strong beliefs. One was that "no one can lie about the future," and the second was that there should be "no secrets in physics." Both concepts were hard to follow but made sense to the complex mind of a brilliant Hungarian physicist. From my view, the notion of accepting deception as standard practice allowed Teller and the Lawrence Livermore National Labs to speculate excessively and postulate on unknown physics. By advocating a policy of no secrets, they could hide behind the cloak of secrecy and say, "If I could only tell you what I can't tell you, then you would believe me."

The story of the X-ray laser has been told many times, but few people realized how strongly Teller believed this was the Third Generation Weapon, following the A-bomb and the H-bomb. The basic concept was to direct the output in a narrow X-ray beam in order to very selectively destroy a target at a distance. It was the very same scenario described by H.G. Wells, Garin, Tesla and Basov, but this would be an X-ray beam of immense power and directionality. Teller was obsessed with the concept, even though the details were a bit fuzzy and left to his colleagues at LLNL. He saw this as the chosen weapon to defeat the Soviet ballistic missile threat, and he found the application and support in Reagan's Star

Wars program. There was one other real problem. Reagan hated nuclear weapons, probably as much as he hated the evil empire, and the White House and all its advisors had no interest in nukes in space.

When Teller finally had a chance to explain the Third Generation Weapon concept to Reagan, he went on and on about this defensive application of nuclear weapons. When the meeting was over and Teller left, Reagan turned to his aide and said in a fake Hungarian accent, "Edvard loves da bomb." There was never any serious interest in the X-ray laser weapon as part of SDI, but eventually I supported the research program because of its potential use as an offensive weapon against our defensive sensors and missiles. We needed to learn if such a system could be real and practical, and I was sure that the Soviets were equally interested in that answer.

Several years later, the head of the Soviet nuclear weapon program, Victor Mikhailov, said such a weapon would be "an evil Jinn," and he called for the end of all nuclear weapon testing.[62] Long after the end of the Soviet Union, and after the scientific basis for even the existence of an X-ray laser was found wanting, the director of the Soviet Arzamas nuclear lab had dinner at my house in Albuquerque. He desperately wanted to know if we really had thought the X-ray laser would work when we had started the SDI program. What he was trying to find out was if it had been a scam from the outset, and he asked an appropriate, although slightly indirect, question. I gave him a technical answer about laser physics development that left him wondering, but I still believed then and now that practical applications had not yet been conceived even given the needed scientific foundation. I believed that the same is true of fusion research, namely that a physics demonstration in the lab does not necessarily mean the thing could ever have a practical application. Teller always assumed that the engineers would somehow figure out how to get the thing to work once it worked in the lab.

When Abe asked me to put together a scientific advisory group for the SDI, there was no question that Teller would have to be included, although the X-ray laser was never an important issue and despite Teller's insistence on ever-increasing levels of funding with few encouraging experimental results. The review meetings were often mostly engineering issues that left Teller bored and provided some time for a nap as the speaker droned on about practical details. It was not uncommon, however, for Teller to suddenly jolt upright and ask the speaker a totally relevant and often utterly defeating question. When Teller was one-on-one with a military person or political decision maker, he did know how to be convincingly relevant and communicate in order to "make the sale." Nobody could be more persuasive than Teller, particularly when he was on the hunt for serious funding for his lab. When we were not in front of others, and when he did not know the answer to one of my many questions, he would say to me, with lots of rolling r's and with his thunderous, gravely, Hungarian accent, "Gerry, I cannot tell you the answer for thrrrree rrreasons. The first rreason, Gerry, is that it is Varry SECRET. The second reason, Gerry, is that it is

VaRRY VaRRRY secret. And the third rrreason, Gerry, is ... (in a quiet voice) ... I don't know. "

It was often the case that Teller did not know the answer about something that he was still imagining and inventing. Nevertheless, he was always curious about the future and optimistic that any critical technical problems could be solved with enough time and money. Getting the money was always of utmost importance. He knew that funding was the indispensable lubricant in the gears of "big science," which was the theme of his laboratory. He also did not hesitate to add whatever political lubrication was needed to unstick the gears of the bureaucracy. In one memorable meeting about the future funding for the X-ray laser, he made it clear that in addition to funding from the DOE, he needed $100 million from the DOD in the next year to keep the program realization from slipping behind by four years. This was at a time in the program evolution that the fundamental physics was still in question, and he was able to focus the conversation on shaving four years off the schedule. He then threatened that he must have the money or he would have to go to the president, who he claimed had already promised these additional funds. Then he did the guilt trip on Abe: "Do you really want to be responsible to force me to go back to the president and say the money is not available from Abe?"

The one person who might have derailed the entire X-ray laser program was the outstanding physicist Hans Bethe. After he went to LLNL and was briefed on the technical details of the program, Bethe wrote me a letter. He had already explained his opposition to the overall SDI program in several public statements, but in the letter he called for continued support of the X-ray laser research effort at LLNL.[63]

Abe could hardly say no to Teller, even though he did not have the funds. The meeting ended with a positive agreement that the money would be found, and indeed the funding was increased in the next year. Eventually, however, the concept was dropped after contradictory technical results from LANL, the archrival of LLNL, but not before large sums were spent.

This was just one of the many instances when Teller would introduce pressure to enhance the position of LLNL over LANL in what should have been a technical competition. I was dragged into such a situation over the ground-based laser because the two labs had two competing concepts. Before it was clear which approach was more suitable, Abe on one of his trips to LLNL apparently said he would go ahead with the LLNL approach. The result was a scuffle that involved the New Mexico congressional delegation. I was asked by Abe to intervene and see that a rigorous technical competition took place before decisions were made. Eventually the LANL concept was chosen and the program went ahead, although it was eventually canceled after a great deal of money was spent. My conclusion from these encounters was the same as Alexi Tolstoi's conclusion in his novel about the mad scientist Garin: "Namely, this smells of higher politics."

CHAPTER 22

Technical troubles

TELLER WANTED US TO PAY for underground nuclear explosions to prove the feasibility of highly focused X-ray beams, while Weinberger and the president emphasized this would be a non-nuclear defense. The chemical laser aficionados wanted us to get on with a ground demonstration of a powerful laser that could be deployed in space and could be coupled to 10-meter optics. We were able to use the laser at White Sands to blow up a section of a Titan booster, which showed some success even though detractors called it a "strapped down chicken test." [64]

It became obvious to all of us in the program that if there was going to be any sort of early deployment, it would involve sensors in space and close-in and long-range missiles deployed on the ground, but we knew these would be ineffective against a determined adversary launching a massive attack. We might introduce some uncertainty in the outcome, but we knew the countermeasures that were already used by the UK in their Chevaline program could be employed to defeat the midcourse and that the ground-based radars needed for the close-in defense could be defeated.

So we persisted with our concept of a multilayered defense, including a space-based layer to intercept the boost phase. Even then we knew there would never be anything approaching a perfect defense, and my belief was that we would have to agree to a cooperative transition to much lower levels of offensive missiles and a greater reliance on a defense that also would have to be cooperative. I remember often saying that the eventual outcome would not only depend on the technology, but even more on the actions taken by the Soviets. My hope was that we could "learn to get along." The president encouraged this kind of thinking by offering to share the results of the SDI with the Soviets.

The idea of sharing our technology with the Soviets was quite counter to our very aggressive efforts in the Pentagon to deny them access to any of our technology and even to mislead them with bogus technology whenever we could. Roald Sagdeev, the head of the Soviet space research programs, once asked me, "How could we share good stuff with the bad guys?" My reply was, "When you become good guys, we then will share."

I gave many speeches explaining the technical issues, but the questions were mostly socio/political/economic. The advocates were anxious to get on with deployment, and they wanted to "deploy now with off-the-shelf technology." I responded that the shelves were bare. They then said, "But when will you be ready?" I would reply that it was too soon to say.

This answer did not satisfy the early deployment advocates, and I could see that we would have to create a credible option before my two-year stay in the Pentagon was over. I believed that ground-based missiles could be effective against a limited attack, but only if we could discriminate against decoys in the midcourse. So this became my major technical focus, and I suggested that directed energy weapons could be used to disturb the light decoys and allow us to tell the real stuff from the decoys, but this meant we still needed directed energy weapons in space. My first choice for midcourse discrimination was the neutral particle beam, but this approach would need a space-based electrical energy supply, and that would probably have to be a small nuclear reactor. Another approach would be space-based chemical lasers, and a few dozen platforms would probably suffice for midcourse discrimination. Either approach was decades away, so I could not see any useful early deployment unless the Soviets cooperated with us.

Given the president's offer to share, I suggested we share the early warning system that I called the Sinai Concept, based on the shared early warning system in the Sinai Desert that had allowed the Egyptians and the Israelis to remove their weapons from the Sinai Desert after the 1973 war. My suggestion was received with lots of skepticism, and one author said I was a naive engineer pretending to be a diplomat. There was no question in my mind that I was not qualified as a diplomat, but I was in the swimming pool without swimming lessons, so I had to learn on the job.

Khrushchev had made the exaggerated claim in 1980 that the Soviets had succeeded in "hitting a fly in outer space" with a missile,[65] but they really had meager results and nobody took that seriously. Quite the contrary, June 10, 1984, the U.S. Army demonstrated a homing interceptor called the Homing Overlay Experiment (HOE) over the Pacific in a test closely watched by the Soviets. The homing sensor or eye on the interceptor was an infrared tracker, and the interceptor maneuvered in front of the on-coming re-entry vehicle with an impact velocity of 20,000 miles per hour at a distance of hundreds of miles from its intended target on the ground. The Army had spent $300 million of their own funds, and many in the U.S. thought it might be just a stunt. Because the intercept took place in outer space but close to the intended target and with no countermeasures such as decoys, I believed it would not be of that much significance. The limitation of this midcourse intercept was that any real attacking re-entry vehicle was going to come as a total surprise instead of a scripted test and was sure to be only part of the attack cloud. There were certain to be attackers along with many decoys and other anti-sensor methods that would make the intercept unlikely.

The SDI program's first technical success, the Homing Overlay Experiment in June 1984, was the result of a Lockheed development supported with $300 million in U.S. Army funds.

By July, I was focusing on boost phase intercept and thinking mostly about space-based chemical lasers that were our most advanced space weapons based on years of effort at DARPA. The Soviets, however, thought HOE was a big deal, and I believe this was the first step in their overreaction to our program. After this success, without our knowledge, they stepped up their secret space battle station program, and at the same time with total openness, Gorbachev launched their diplomatic counteroffensive to put a stop to our program.

One day, I was asked to show up on a Saturday morning for a private meeting with the secretary of state, George Shultz. I showed up at the State Department and was escorted to the breath-taking splendor of the seventh floor, filled with antique furniture and the office of the secretary. Much to my surprise, the head of the CIA, Bill Casey, and two other very prominent scientists, Jonny Foster and Bud Whelan, also were there. Casey was bald, mumbled a lot and was wearing bright green pants and a brilliant yellow polo shirt. I was sure he stopped off on his way to the golf course. It was clear that he wanted to get on with the meeting so he could do more important things, so Shultz started by asking me to explain the technical foundation of the SDI. I did not waste a lot of time going into detail but said it was a research program and that the outcome was uncertain. That was almost the end of the play, and I later decided that the script was already written with me making unsupportable claims that Whelan and Foster would then defeat so that Shultz could prove to Casey that the SDI was bogus. More importantly, he would have strong ammunition to use in confrontations with his long-term rival, Caspar Weinberger. I could see that the program was not about technical issues, but about beliefs, perceptions and at some level, internal rivalries.

There were many opportunities to debate outstanding scientific leaders and political specialists who did not have scientific backgrounds, and I never had a problem with the technical critics, because I often agreed with them that there were many technical unknowns. We would then disagree over which were the best approaches and how much should be spent, and their problems boiled down to opportunity costs. Namely, if we spent too much money on SDI research, there would be less money for their favorite and far more important research. The computer scientists believed the biggest problems were in the software and that there were not enough computer scientists to go around. So our arguments were over degree rather than the fundamental notion of doing research on ballistic missile defense.

Whenever I gave public speeches on the subject, there would always be early deployment advocates in the audience, and they would take me to task. I would only say that the Fletcher study intentionally de-emphasized any short-term or narrowly defined program elements and would work toward a jointly managed transition in cooperation with the Soviets. The reaction of the early deployment advocates was, "How dare you carry out a defense program that depends for its success on what the adversary does? Do you expect us to pay for your research and give veto power to the Soviets?" I would then respond with a quote from Sun Tzu's classic book *The Art of War,* written 2,500 years ago, "To secure ourselves against defeat lies in our own hands, but the opportunity of defeating the enemy is provided by the enemy himself."[66]

CHAPTER 23

Budget battles

THE BUDGET AND THE POSSIBILITY of cuts were the subjects that dominated almost every discussion during that time as we strategized ways to fend off cuts in FY 86. We were pushing for an SDI FY budget of $3.8 billion, a tremendous increase, and the rest of the Pentagon was looking at substantial cuts as a result of the 1985 Gramm-Rudman balanced budget act. Decisions would have to be made about setting priorities, and Weinberger had made it clear to everybody that SDI was job one. Other programs would have to take cuts, but we were to fight for the $3.8 billion for SDI. If we were successful in arms control reductions in offense, the "pay now" would give us leverage in negotiations to allow us to pay less later. Unfortunately, our financial credibility was beginning to slip because of the slow start of programs and our inability to deliver on our rhetoric of instant miracles. We were clearly under the gun to get deliverables rather than do research. I knew we could not pay for everything we wanted, and we still had to worry about Senator Proxmire with his

Golden Fleece Award, the $100 hammers, the $2,000 coffee pot and the Navy's $600 toilet seat.

By April it looked like our budget would be "only" $3.5 billion, and it was clear we needed to show some results or at least show we could spend the money. Unfortunately, we did not have an up-to-date accounting of our outlays and we were not in a good position to argue with the services about priorities, cuts and commitments. The DOE was pressing for more and more funding so they could accelerate the X-ray laser program, which was now called Excalibur. Evidently Edward Teller had been describing new concepts involving hundreds of beams and demanding hundreds of millions more dollars. Meanwhile, the House of Representatives was talking about $2.5 billion, and others were suggesting we settle for $2.9 billion. I was an observer in the budget fights, not a participant, but I did not see much harm in slowing down and focusing on the real technology issues such as midcourse discrimination.

By May 9, our arguments were unraveling and our support in the Senate was crumbling for all defense spending. Nevertheless, we were holding out for a giant budget increase. By the end of May, I was convinced we had more than enough money to execute a research program to resolve critical issues. I was opposed to dumb demonstrations and suggested to Abe that we focus on "drop-dead demos" that would allow us to cut back on the obvious losers. The waste, fraud and abuse accusations were building up daily to batter the defense budget, and I was prepared to argue for $2.5 billion while trying to defend the president's pledge to share our results with the Soviets. Many programs were taking deep cuts, while we were growing our investments so we could give away our advantages. Our hope was that the Soviets would agree to accept our generosity and, in return, cut back in their offensive investments.

By July, the pressure was building to appoint a suitable advisory committee and form a subsidiary that could more comprehensively deal with all the political, economic and systems technology issues. It gave me no pleasure knowing that I had suggested this to Abe almost from the beginning of my assignment one year earlier. Abe's senior advisors – Ben Schriever, John McLucas, Jim Fletcher, John Foster and Simon Ramo – knew we needed help and formed an informal committee to make recommendations about how to move forward. They all believed we needed a high–level, quasi-governmental support organization very much like the other nonprofit corporations that existed to support the Pentagon and the services. Schriever suggested the new organization should employ civilians with the highest technical credentials who could help with technical evaluation, systems analysis, policy and arms control issues, procurement planning, etc. I was really excited about this idea and felt that one of the most useful roles of such an organization would be to interface with the media, who never seemed to be able to deal with the contradictions, hyperbole, confusion and technical realities when they were interacting with so many sources from the services, industry, Congress and the administration. A unified and balanced story would be helpful to all concerned.

Abe asked me to start working on a concept paper for this new not-for-profit corporation to support the SDI, and he sent a memo containing a concept of a new organization to the appropriate Pentagon offices for their feedback. The senior group stated the following: "SDI is technically and management-wise the most complex and difficult program ever undertaken by this country – perhaps by an order of magnitude. It is also the most important for the survival of the free world."[67] This group certainly had the credentials to make such an evaluation since Schriever and Ramo had been at the heart of the U.S. creation of its ballistic missile capabilities that blunted the Soviet missile initiatives three decades earlier. I agreed that the program was complex, but from my view, this complexity did not stem from the technical or management aspects, but because so much depended on largely unpredictable aspects of human behavior. Nevertheless, I was pleased when the senior advisors recommended the formation of the SDI Corporation, a not-for-profit organization very much like the federally funded research and development corporations, Federally Funded R&D Centers, or FFRDC, which already existed to support the DOD and the DOE.

Colin Powell, knowing the importance of this new FFRDC to Weinberger, loaned us an Army colonel named Alfonso Lenhardt to serve as a management consultant to help us deal with our problems with administrative decisions. Lenhardt needed a temporary job before his next assignment in Germany, and I was sure Powell had big plans for the bright and charming military officer. Lenhardt quickly grasped the importance of the task and knew how to get the ball rolling. With my help, he drafted a proposal that Abe took forward in September 1985.

Although it had support from some of the most experienced and wisest members of the senior defense community, the proposal met an immediate backlash. The services hated the idea because they said it would just strengthen the decision-making capability of the central organization and reduce their highly cherished independence. Brigadier General Bob Rankine, who preceded me as the deputy director of the program, said this would be "a further erosion of the fundamental management principle under which the SDI program was formulated: decentralized execution by the services." Of course he was correct, because that was Abe's intention all along.

The so-called "beltway bandits" (or as I preferred to call them – the "highway helpers") hated the proposal because it would reduce the government need for their support. They told Abe, "Large numbers of qualified professional service firms stand ready to perform the needed services on a competitively acquired, independent, conflict-free and accountable basis." They did not want to see a not-for-profit support contractor in direct competition with the for-profit support contractors. In January 1986, progress on the proposal was not forthcoming and the senior group went to Weinberger to emphasize the need to move ahead with the concept. Weinberger then asked the SDI advisory committee to take on the task of getting the new FFRDC off the ground, but this just bounced the ball around even more since the advocates of a new FFRDC were facing a force stronger than any other in Washington, namely the source of all appropriations, the Congress. The Congressional representatives had doubts about creating a permanent SDI institution since they were not convinced that the U.S. really needed ballistic missile defense, and several of the decision-makers on the other side of the aisle guessed the whole thing would evaporate in a few years. As a result, the concept was shelved in a few months. What did grow from these seeds was an independent analysis support group staffed by senior people from government labs and think tanks. It seemed that only the Soviets viewed the SDI as a vital part of U.S. national security that rivaled the Manhattan Project in importance.

One of the best aspects of developing the FFRDC proposal was working with Colonel Lenhardt, who shared my sense of humor and viewpoint on military culture and the irony of our jobs. Our office reported to the secretary of defense, but when you entered the outer office where the staff sat, you immediately knew you were in Air Force land and the culture was Air Force culture. Because I had come from civilian life directly into an Air Force two-star job, I did not understand the differences between the services. An Air Force general that I sat next to at a banquet generously explained these differences to me with the following story: "An army platoon is ordered to secure a facility, so they surround it and hold it until the occupants surrender. A Marine group is given the same orders, so they fix bayonets and charge the building, killing the occupants, climb to the roof, wave the Stars and Stripes and cheer 'Semper fi.' The Air Force squad given the same orders makes a few phone calls and negotiates a lease with an offer to buy."

I observed this interservice culture clash almost daily as Air Force colonels would casually stroll into the office and, seeing Lenhardt, an Army colonel in the support area, begin to harass him about some trivial issue such as travel orders. Rather than explain his job, the Army colonel would go into an eye-rolling, foot-shuffling routine that would just cause even more hassling, so, hearing the commotion, I would walk out of my office and, with obvious sarcasm, accuse Lenhardt of disrespect and insubordination. The Air Force colonels immediately knew they were engulfed in our private joke at their expense and would exit as fast as they could, leaving Al and me in convulsions of laughter.

CHAPTER 24

Adventures in public relations

Abrahamson personalizes a Washingtoon cartoon with his signature smiley face.

ONE OF MY MOST DEMANDING roles was representing the SDI program in the face of fierce opposition from many of the most impressive members of the scientific community. We had many virtual "face-to-face" debates that took place in published articles. One of the first appeared in the June 1985 *Physics Today* magazine, published by the American Institute of Physics.[68] The opposing view was from Wolfgang Panofsky, who was the director emeritus and professor at the Stanford Linear Accelerator Center and a former president of the American Physical Society. Although I was never president of the society, I was an active member and also received the honor of being chosen as a fellow for my work on particle beam fusion. Nevertheless, Panofsky had the technical credentials to attack the program. Although he could have attacked the details, his most telling statements were more general:

> Mark Twain said, "There is something fascinating about Science. One gets such wholesale returns of conjecture out of such trifling investment of fact." I could add to this, the SDI is truly impressive; one can base so much political and strategic posturing on such limited and technical and military potential … the present program may threaten to fuel the arms race, violate the ABM treaty, and upset strategic stability. …At best, a premature demonstration hypes the impractical; to quote Samuel Johnson, it is "… like a dog walking on his hind legs. It is not done well

but you are surprised to find it done at all." ...The media have dedicated much space to the debate over whether or not it could work ... it is unclear what mission 'It' is to accomplish ... Soviets accuse SDI of leading inevitably to the weaponization of space and signaling the beginning of a new high-technology arms race. ...even a small fraction of the world's nuclear weapons can end civilization.

My article described the SDI research program and the technical issues that were unresolved, but I had little to disagree with Panofsky. The one major difference was that I left as a vital but unanswered question the issue of how the Soviets would respond. I said that there would have to be a jointly managed transition to incrementally introduce balanced defenses along with a reduction in offensive missiles. I shared Panofsky's concerns about the issues of arms race stability, but also strategic stability. Facing the realistic possibility of eventual mutual annihilation was not a desirable outcome by any means. Panofsky never considered whether the Soviets might be prepared to negotiate the proposed managed transition to the Dyson "live-and-let-live world." He also never considered the possibility that the Soviet Union itself might not have an unlimited future.

Another debate erupted when I was asked to comment on Ashton Carter's 1984 paper for the office of technology assessment of the Congress that concluded, "The prospect that emerging Star Wars technologies, when further developed, will provide a perfect or near perfect defense ... is so remote that it should not serve as the basis of public expectations."[69] I agreed with Carter, which disappointed my SDI colleagues. I found little to be wrong with his technical analysis, and I was pleased that he found little to be wrong with my *Physics Today* article. He sent me a personal letter complimenting me for my "technically judicious article" as compared to some of my colleagues in the SDI who tended to be less careful with their arguments.

The published debates continued, and the next published pro and con was in the September issue of *Science Digest* magazine.[70] The magazine hired an artist to draw a picture of me depicted against a background of mythological space warriors, and this drawing was splashed across the cover that appeared on magazine newsstands. (Once, when I was asked for identification while writing a check at the grocery store, I reached over and used the magazine cover.) In this article, the opposing view was from my former quantum mechanics professor from Cornell, Hans Bethe. His reputation in physics was without parallel, and he had received the Nobel Prize for his pioneering work in physics. Bethe attacked the technical issues in a comprehensive manner and one by one pointed out all the problems, which incidentally I had described in my *Physics Today* article in June, which he obviously had not read. He argued that we could never have a perfect defense. His most convincing argument was the difficulty of developing and testing the battle management system. He concluded that "the entire system could never be tested under circumstances that were remotely realistic." His conclusion was that "mutual deterrence is all we have and all we can have." In my article, I called for creativity associated with the exploration of a new field.

From the first day on the job, I became the "go-to guy" to interact with the many audiences for program explanations. I gave dozens of speeches at the rate of at least one a week, and I found myself sent by the Pentagon public relations folks to address hostile academic audiences: Caltech, the University of Chicago, Cornell, MIT, etc. One day it occurred to me that Abe seemed to always be giving talks south of the Mason/Dixon line at places like Georgia Tech and the University of Tennessee when I was entertaining the academics in the North. I suggested that Abe give me a chance to try my speaking skills on more favorably inclined audiences and he said OK, you talk to a German military group in Munich and the Disabled American Veterans in Jacksonville, Mississippi. I managed to get the DOD to allow me to use their international communication tools to stay in the Pentagon and give my talk with a photo of me on the dais in Germany. After the talk, the facilitator said I got a standing ovation. I suppose it might have been true, but I was looking forward to heading south and addressing a live and agreeable audience.

In Jacksonville, I was treated with a great deal of respect and treated to a splendid lunch of fried catfish, hush puppies and some sort of green stuff I think they called greens, followed by hot apple pie with a dollop of fresh vanilla ice cream on top. It was time for my speech, and I was introduced by a retired general who made me sound like a hero. I was pumped up to give the talk of my career. As I started to talk, I was interrupted frequently by calls such as "Yes, brother," and "God bless you." This time I could see and hear the warmth of the audience response as they stood and applauded. Filled with pride and confidence, I wanted to continue to enjoy the support, so I asked if there were any questions. An elderly gentleman jumped up to catch my attention. He then went on to thank me for "defending the U.S. against alien invasions from outer space." I was speechless for a moment, then regrouped and replied that it was late and I had to catch a plane. On the way home I decided that I was better suited to being burned in effigy at the University of Chicago. I did not ask to head south again.

I kept giving talks and decided that at least half of my job was being the official program spokesperson. As the speaking engagements continued, I found myself straying more and more into strategy and away from the weak technical merits of our story. It became more likely that the high-power lasers and particle beams would be less important in the long term as Dyson had suggested in his "live-and-let-live policy." Such strategic thoughts were hardly ever discussed in the Pentagon. The typical time horizon was hours and days and seldom even stretched into months. It seemed that only the president thought about big ideas and grand visions, with no clear path or the slightest details, and everybody else scrambled around what to do that required complexity and thought. Every day was filled with "rush jobs" with no time to consider alternate hypotheses or long-term implications. Many years later, Bud McFarlane, the president's national security advisor, commented that the SDI was really a cleverly developed sting operation,[71] but the carrying out of any complex secret plot was inconceivable to me. Everything just happened with no consideration for the long term implications. How could anybody be concerned about the MAD doctrine when you had to think about and explain a $600 toilet seat?

Questions about government spending become the focus of a Bloom County comic strip.

CHAPTER 25

Smart rocks and brilliant pebbles

In addition to public relations, my most immediate task was to get on with review and oversight of critical programs involving surveillance, battle management and weapons. I knew that if you can't find targets and can't manage the decisions about shooting at them there was no need for weapons, but all the publicity and attention went to the weapons. Jay Keyworth, the president's science advisor, was concerned that any weapons in space would become vulnerable targets to Anti-Satellite (ASAT) weapons. His focus was on developing lasers on the ground that could deliver the killing blows to boosters by reflecting the beams off mirrors in space. He thought we could develop methods to make the mirrors survive and that would be much easier than making the self-contained space battle stations survive. Furthermore, there "would not be weapons in space, but only some flimsy mirrors" that could be replaced with existing boosters.[72] Of course, we had no proven concepts for ground-based lasers, nor lightweight replaceable space mirrors, and we already had a large and well developed DARPA chemical laser weapon program with substantial political backing from Senator Wallop and the aerospace industry. There was also a well-established community of airborne laser enthusiasts in the Air Force, but my panel in the Fletcher study had already decided not to fund that because the laser-carrying aircraft would have to fly permanently over the enemy territory, waiting to be shot down. So at least we were able to put a stop to that program … or at least I thought.

Then there were also the X-ray laser concepts at LLNL and LANL and the constant marketing from Edward Teller and his sidekick, Lowell Wood. I had already decided that we should explore the feasibility of X-ray lasers, not because I thought they could be used to defend us, but I thought that if the Soviets had such weapons, they could attack our defenses in space. Furthermore, Hans Bethe, one of the strongest SDI detractors, sent me a personal letter saying, "The X-ray laser should not be disqualified simply because it exploits a nuclear explosion as a power source."[73] So with both Teller and Bethe supporting the X-ray laser, I agreed we had best fund it adequately. As it turned out, the pressure from Teller was more than adequate to get support from Abe.

I also knew from the many years of prior work and the "brilliant success" of the HOE demonstration that we could intercept re-entry vehicles (RVs) in the midcourse if we could figure out how to separate the real RVs from the decoys that looked like RVs, as well as the RVs that looked like decoys. We called this the midcourse discrimination problem, and it looked to me to be the key to any successful early deployment. So we had lots of weapon issues that needed a lot of money. My priority list included dealing with the following issues: 1. midcourse discrimination, 2. feasibility of ground-based lasers 3. survivability of space weapons, 4. effectiveness of all weapon concepts and 5. cost trade-offs. All the early deployment concepts were way down on my priority list.

I had suggested to several potential contractors that they consider "smart rocks," which were small homing missiles, but Lowell Wood could go one better on this and proposed brilliant pebbles. Of course, I countered with "sagacious sand" that could be directed into the path of an attacking swarm of RVs and decoys and would shred the lightweight decoys. This might even have turned into a workable concept, but there were many distractions and funding issues to prevent us from exploring any really new ideas.

Other complications arose when we considered human limitations and software problems. Although for most of my career, I was primarily interested in weapons delivering energy at the speed of light, I always wondered about the human component. I had a vision of two space battle stations with the ability to track, identify and strike distant targets, but with human decision-makers sitting at the wheel asking, "Should I shoot, or should I wait until I have more information?" When I became involved in the Fletcher study and it became obvious that an effective defense would have to strike at boosters shortly after they are launched, I asked an Air Force general, "Who would make the decision to start the defensive response?" The answer was, "Of course it would have to be the president."

The next obvious question was how he could make that decision in seconds or at most minutes, and the answer was that he would have an aide by his side with "the football." The president could decide to start shooting defensive weapons instead of just striking back in retaliation .... or he could wait a bit longer to see what was happening. "What would happen if he were sleeping or otherwise physically occupied?" I got the same answer about the aide and the football. So I then said, "A first strike might be limited, staggered to exhaust the defenses, faked or real, and how could there be enough instantaneous valid information for a decision?" The general responded by saying he had an appointment. This interchange left me convinced that much of the defensive response would have to be automated, but how could we trust the computer software to make such decisions?

When I became more involved with the defensive system architecture involving at least three different but coordinated defense layers, as well as sensors on the ground, in the air and in space, I became more concerned about humans getting involved in managing this complex process that might last for only tens of minutes. The various components had to "talk to each other" as well as talk to the humans managing the battle as it unfolded in unpredictable ways. Again I became convinced that humans could make long-term decisions, but then they would have to trigger computer-managed actions, and they would have to trust the hardware and software.

As the concept of the multilayered, coordinated, sensor-driven system evolved, the question of realistically testing the system software kept coming into our discussions. Eventually I started to believe naively that we could develop error-free software as long as we could employ only software engineers who were trustworthy. Because we planned on a procurement involving many defense contractor suppliers, I then became more and more worried about the competence and security issues. Some estimates were that there could be thousands of software engineers. How could they be trusted not to introduce a hidden bug that would appear only seconds after the war started?

I was soon corrected by the experts to stop talking about "error-free" and talk about only "error-tolerant" software. That sounded a lot better, so I asked for examples of existing large, complex systems that were resilient and able to deal with infrequent but important errors. I was never satisfied with the answers that sounded a bit like, "This is an important research area that requires substantial funding," so I asked who was really good at testing software to find problems. Invariably the answer came back to the fact that there had to be humans in the loop, which was where I started this quest.

I asked to talk to the experts. I was told that the right place was the Royal Signals and Radar Establishment in Malvern, England, so I decided to go there on my next trip involving the development of cooperative programs with the U.K. I was particularly impressed by their Chevaline program for strategic nuclear-tipped missiles as a deterrent against the Soviet Union. This program that was managed by Stan Orman included convincing countermeasures that we could expect and would have to be overcome. So the U.K. offered both the model for the opposing force and maybe the way to test software so it could be trusted.

Malvern is a bucolic town in central England. The facility looked like it was right out of a World War II movie. When I got there, they graciously put me up in a quaint, one-room brick cottage on the lab grounds. It was one of those damp, cold fall nights in England, and I had meetings scheduled for first thing in the morning, so I went to bed early, noticing that there was no heat in the cottage. A couple of hours later, the temperature had dropped about 20 degrees, so I piled everything I could find on top of me and tried to sleep, but to no avail, and I had to get up early. I searched the room and found a giant electric tea kettle, which I filled up, turned on and after the warming steam filled the room, I dozed off. As soon as the water had boiled off, I had to fill it again, and again, and so I survived the night. Early the next morning, when I turned on the light, I was shocked to discover that the wallpaper had been steamed off the walls and was piled in neat folds onto the floor. I never said anything to my hosts. I am sure that after I left, they figured I had been doing some sort of a strange and freaky cult-like activity in their cottage, and they probably are still trying to figure out what happened.

I then went off to learn about testing software, but by then I had direct evidence that the unexpected can happen, and my questions that day were all about dealing with freak occurrences. There were few answers since they really did not know how to test software in the face of predictable environments, let alone in circumstances involving complete surprises.

While I was stewing about reliability of software in 1985, Abe asked a group of computer experts to look at the problem, and their work boiled up into a controversy within the computer community. One member of the advisory group, David Parnas, resigned from the study and claimed that the system could never be tested and could not be trusted, and its success is "less likely than 10 thousand monkeys randomly typing the Encyclopedia Britannica." Danny Cohen, the leader of Abe's study group, argued in a public debate in the fall of 1985 that a resilient system, like the telephone system, would still function even though "individual components don't work."[74] Prompted by Cohen's remarks, I went to an old friend, Sol Buschbaum of AT&T, and he supported the argument that – with the right kind of system design, testing and redundancy – the system could be made to work. So when I was asked questions about this, I frequently used the telephone system as a proof of principle that we could do it.

As I continued to study the issues of reliable and secure information systems, it became increasingly clear that both communication and power distribution systems would occasionally fail with major losses of service to millions of customers. The problem was often one of those freak occurrences that nobody had considered, and after the fact, the problem was corrected with enough resiliency and backup so that it could not happen again ... until something else happened. The reliability of complex systems with many non-linear, interdependent variables will always be unpredictable. The only thing we can predict is that surprises will always happen. I was faced with the dilemma of figuring out how to effectively deal with very unlikely but high-consequence surprises. I never did figure that one out, except to try to avoid such situations, and also to prepare to return to normalcy after the inevitable failures ... that is, if we survived. Of course, an even worse problem would be any hidden faults in the software intentionally inserted by an adversary working on the inside of our program. These faults could lie there dormant until the moment of the attack. I never did figure out how we might protect the software being developed by dozens of contractors and hundreds or even thousands of software engineers.

CHAPTER 26

Back in the USSR

As I struggled with these dilemmas, I was also trying to gain a better understanding of what was actually going on in the USSR. What were the Soviets thinking, and how much did they actually know about what was happening in the U.S.?

On March 11, 1985, Gorbachev was elected to the position of general secretary of the Communist Party of the Soviet Union. His main goal was to revive the Soviet economy by reforming the economic and political systems. Gorbachev also wanted to pursue nuclear arms reduction, but the Soviet military remained concerned about keeping up with weapons developments in the U.S., and when he and Reagan met for the first arms summit in 1985 in Geneva, negotiations broke down over Reagan's commitment to the SDI.

I was not aware at the time that the Soviet military had been fed bogus technical intelligence from Project Farewell, a technical disinformation program engineered by the CIA. In the early 1980s, the CIA had turned the KGB official,

Colonel Vladimir Vetrov, who was in charge of evaluating U.S. technical intelligence collected by the KGB. Vetrov provided a convenient channel for the CIA to "help" the Soviets with "misleading information pertinent to stealth aircraft, space defense and tactical aircraft."[75] One example of concrete evidence of the success of this disinformation campaign was the bogus Siberian pipeline software that resulted in a June 1982 "three-kiloton" explosion detected in Siberia by U.S. satellites. According to Tom Reed, former secretary of the Air Force, "We were blowing up Soviet pipelines, infiltrating their computers and software …. inflicting their industry with bugs, viruses and Trojan horses."[76]

Also, unbeknownst to me at the time, the Soviets were far ahead of America in successfully launching a demonstration test of a space weapon, which would have been the key to their ability to destroy our space assets if they had pressed on with development and deployment. This had been predicted by some in the U.S. defense community, including Steve Meyer, an MIT Soviet expert, who said, "If we push ahead with SDI, the Soviets will be the first to demonstrate a laser in space … it will not have any military utility, it will not be a space weapon. The intent will be for political strategic impact."[77] Ashton Carter speculated on a purely "fanciful" example of a hypothetical Soviet effort, saying, "The Soviets launch a space structure combining a moderately powerful chemical laser with a large mirror … test destruction of satellite targets … eventually a total of 10 such lasers appear in orbit." He also pointed out in 1986 that an "ASAT attack on crucial sensors based in space is probably the cheapest and most effective offensive countermeasure."[78] A comprehensive history of the Soviet Union's giant Terra laser program published in the '90s convinced me that the Soviets had the experience and desire to commit to very large investments in laser weapons if they had the political will to do so.[79]

Much like here in the United States, Soviet scientists and military advisors saw the arms race as a way to leverage funds for their own pet projects. M.I. Gerasev, from the Soviet Institute for the USA and Canada, said, "We had plenty of zealots who greeted Reagan's SDI with open arms. They came running with comprehensive projects expecting to be showered with funds."[80] Evgeny Velikhov, the leading scientific arms control advisor in the Soviet Union, who accompanied Gorbachev to all meetings with Reagan, wrote in his memoir, "Our negotiators and military experts were convinced that by 1990 the Americans would deploy space weaponry."[81] I assume that Velikhov would have known that this was nonsense. The military advisors told Gorbachev that the U.S. had developments including "kinetic energy nuclear weapons, in which a nuclear explosion creates a stream of metallic fragments of small mass that travel at more than 10 kilometers per second and are capable of striking targets in space, including warheads, a direct hit."[82] The physics behind such a claim escapes me, but of course, I had no say in the matter. Gorbachev had also been told that full-scale development of "X-ray laser weapons, directed electromagnetic radiation weapons and kinetic energy weapons is expected to occur in the second half of the 1990s."[83]

Not only had Gorbachev been told that the Soviet scientists had proven that a space-based missile killer was possible, but "the U.S. has achieved results in this area which surpass those of our country."[84] The Soviet leader had been warned about "space-based systems with ... full-scale deployment expected after 2010." They said, "Overall the Soviet Union lags approximately four to five years behind the U.S. in research on creating the elements of a space-based missile defense echelon,"[85] and "Americans think that a multi-echelon missile defense system should allow, at most, 0.1 percent of the attacking missiles to get through."[86] If I had been told such a totally ridiculous projection from one of many industrial marketers who hounded me, I would have banished them from the Pentagon. But what was behind such claims? Was it the marketers of the Soviet military industrial complex taking advantage of the technically incompetent decision-maker, or did the U.S. have some hand in creating this perception?

The head of the Soviet military industrial complex, Oleg Baklanov, had already decided that Gorbachev was a liberal, non-confrontational, lead-from-behind diplomat who did not understand nor appreciate science and technology. Baklanov tried to convince Gorbachev to launch a space race, which he argued they could win, but Gorbachev was more worried about the Soviet economy that was collapsing around him. Gorbachev did not help his economic revolution very much by instituting prohibition, which cut off one of their important sources of income, selling vodka. Gorbachev thought vodka was creating a lazy and drunk workforce and making their country unable to compete with the rest of the world. Meanwhile, the Soviet military and their military industrial complex were spending about half of the national budget despite the USSR's many impoverished client states such as Cuba, Vietnam and Eastern Europe.

I am sure many in the Soviet military industrial complex saw SDI as an opportunity for funds, and Gorbachev must have been acutely aware of the rising tide of pressure to invest in space weapons. As it now appears; however, Gorbachev was not influenced by technical arguments that he barely understood. According to Baklanov, Gorbachev "had a poor grasp of the subject matter. ... no understanding of it, no defined ideas about the issues of defense."[87]

Many years later, Victor Mikhailov, who became the head of all Soviet things nuclear as head of the Russian Federation of Nuclear Energy, wrote in his memoirs that Gorbachev was in no mood to listen to his technical advisors, but instead, "Gorbachev has to take the blame for the attempts to demolish the military industrial complex. He almost ordered that the directors of our enterprises be squashed, treating the talented scientists and organizers like bedbugs."[88]

As Gorbachev struggled to implement his new policies of glasnost and perestroika and make sense of the dire warnings from his military advisors, he faced a major national crisis. A catastrophic nuclear accident occured in Chernobyl in 1986. The Chernobyl disaster is known as the worst nuclear power plant accident – both in economic impact and casualties. The catastrophic accident, due to carelessness and a bad design, resulted in a steam explosion that blew off the 1,000-ton lid over the reactor and led to a release of radioactivity that caused 28 short-term deaths and thousands of cases of cancer. As a result, more than 300,000 people were resettled. Historians claim the disaster and the way the Soviet government handled it contributed to the continuing decline of confidence the Soviet people had in their government. The accident also deeply impacted Gorbachev, who wrote in 2006, "Chernobyl opened my eyes like nothing else: it showed the horrible consequences of nuclear power ... One could now imagine much more clearly what might have happened if a nuclear bomb exploded."[89]

A few months later, in September, a well-known Soviet cruise ship, the Admiral Nakhimov, filled with wealthy vacationers, collided with a freighter in the Black Sea in perfect weather and sank rapidly. Hundreds of lives were lost. This tragedy caught the attention of most Soviets and became a national "source of pessimism and foreboding."[90] Then on Oct. 3, 1986, an accident led to the sinking of the K-219 ballistic missile submarine with its two nuclear reactors, 16 missiles and 34 nuclear warheads within 1,000 kilometers of Bermuda. The Soviets claimed that the accident was a result of a collision with a trailing U.S. submarine, but America denied that. Strangely, the unanswered question of the fate of the nuclear missiles is confounded by an unfounded rumor that two years later a Soviet research ship inspected the wreck and found the launch tubes open and the missiles gone. The bottom line is that Gorbachev probably had lost whatever faith he had in the Soviet Union's abilities as a modern technological force in the world, and he could only turn to his hope of fundamentally changing his economy.

Meanwhile, the United States was facing a crisis of its own after the loss of the space shuttle Challenger, which exploded just after liftoff in January 1986, killing seven, shocking the nation and delaying space-based experiments scheduled for the SDI. Gorbachev sent a telegram to Reagan expressing his grief over the tragedy; however, the *Los Angeles Times* reported, "Gorbachev today expressed sorrow over the Challenger shuttle disaster, but Polish and Czechoslovak papers said the seven astronauts died in a 'Star Wars' program aimed at military domination." [91]

CHAPTER 27

Reaching out to our allies and the VP

ONE OF SECRETARY OF DEFENSE Caspar Weinberger's goals was to gain support for the SDI from our allied nations. His plan was to offer contractors in those countries a portion of the SDI funding in order to lay the groundwork for an international strategic defense effort. As a result of Weinberger's international thrust, I was invited to weekly parties at embassies, visited by foreign dignitaries and sent on many junkets to foreign capitals. The result was sometimes humorous, sometimes ludicrous and sometimes dangerous.

Our first attempt to involve our allies in the research program was with the French. Fortunately, I knew several French scientists who were experts in laser technology. We established a tentative area of joint work before Reagan visited the president of France, Francois Mitterrand, to get a signature on the bottom line of the agreement. We were on the verge of our first real allied agreement when Reagan used the term "sub-contractor." Mitterrand's reaction to being called "le sous contracteur" rather than a full-fledged partner did not go over very well. Suddenly, the deal was off. Mitterrand offered to start his own high-tech program in Europe with France as the lead, not the sub. The French program was called Eureka. A few weeks later I was stunned to see a large moving van outside my office with the name *Eureka* emblazoned upon it. I told Abe the French had arrived.

The joint efforts were not building up momentum, so the president asked Vice President George H.W. Bush to go on over and get signatures on more agreements. Bush needed technical backup, and his assistant called and asked me to help out by accompanying the vice president on the trip to Europe. I hesitated at the thought of upstaging my boss and asked if it would be more prudent for Abe to go instead of me. His assistant said they had intentionally not asked Abe because they wanted a low-key, backseat guy and not an impressive general who might draw attention away from the vice president. So I agreed to go with the vice president as long as Abe asked me first. The next day Abe did this with a fake Jewish mother approach about not worrying about leaving him behind by himself all alone in the dark.

I was invited to brief the vice president in his office, and I remember how gracious he was even though I was a bit late because of a scheduling mess-up. I noticed his unpretentious demeanor, his plaid watchband, his signed photo of vice presidential candidate Gerry Ferraro on his coffee table. We sat in the overstuffed armchairs and chatted like old friends. He was a good listener, understood my comments and asked that I meet with a small group at his house for cocktails and dinner to talk about some of the issues before the trip.

I was tremendously excited and determined to be on time for this appointment. I made sure that all arrangements were made for a car to go from the Pentagon to the vice president's mansion with plenty of time to spare but, as I was leaving my office, my secretary Judy told me my underwear was showing through a very large hole in my suit pants. I had been working day after day for long hours and had not realized that I had worn out my only suit. My house was out of the way, so having been late the first time with Bush and unwilling to be late again, I decided that the jacket would have to cover the hole. I spent the entire time at the cocktails before dinner standing awkwardly and pulling my suit jacket down as much as I could to cover the hole. Everything seemed to be working out, but Charles Krauthammer, a reporter, was there … in a wheelchair. His position in the wheelchair gave him a bull's-eye view of the hole in my pants. When I noticed that he noticed, I blushed and begged him with my eyes not to tell anyone. Fortunately, he didn't say anything.

Following dinner, we had a roundtable discussion. The half dozen participants all were notable members of the strategic/military/political establishment. They spoke of the "ambiguities and ambivalences" that might conceivably lead to advances in arms control. The group members had little interest in discussing any of the technical issues and the reality of the R&D program. They had no enthusiasm even to consider "what is it" or "will it work." They were even less interested in what it would cost but were very worried about the possibility of the SDI program creating problems within our alliances.

Krauthammer seemed the most enthusiastic about SDI since it "could become a real bargaining chip …. possibly space strike weapons." Somehow he knew that this aspect of technology was of primary concern to the Soviets. I agreed with Krauthammer, that is, if we had something convincing to trade. As it turned out, their military was already convinced.

The vice president listened without comment to all of this back and forth about what the former Secretary of Defense Harold Brown called "not a very good idea …. a mistaken commitment to a real gamble." As I listened, I wondered how Bush would react to all this confusion, and he then calmly closed the discussion with a brief comment: "This is a listening trip … a prudential step…not an SDI trip." He obviously did not share Reagan's enthusiasm, but I was to see how he would deal with other world leaders on the subject. One decade later, I learned from Dana Carvey on *Saturday Night Live* that the comment "Wouldn't be prudent at this juncture'" was the hallmark of President George H.W. Bush.

The dinner and discussion had gone well, and I was looking forward to joining the vice president on Air Force II. In preparation for the impending trip to Europe, I rushed out and bought a new suit to get ready for what I thought would be the high point of my new career as an international techno-diplomat. Again, events out of my control would change my plans. A terrorist incident in Europe caused Reagan to refocus Bush's trip from SDI to counterterrorism, and my involvement was abruptly canceled at the last minute by a phone call from Bush's assistant. I was totally crushed and left with the task of telling Abe about my misfortune. He did his best to appear understanding and sympathetic, but it was hard for him to hide the slight grin on his face.

One of my most memorable foreign entanglements took place at the British embassy, holding a glass of wine and casually responding to questions. I remember being asked, "Would you provide us with 150 million." I did not know how to respond. Was that dollars or pounds? Was this amount requested for one year or for many years? I said in a less than serious way, "Is that all you want?" I don't think they understood my sense of humor, and one of them looked very upset, turned very pale and ran from the group. I always figured that he thought they had asked for too little and he was calling the boss to get new instructions.

I also recall when Abe asked me to have a short and informal talk with the Japanese foreign minister, whose name also was Abe but pronounced differently. I said I would go over to the embassy and talk to him, but Abe said I would have to go to Tokyo for this brief chat. Naturally, I made arrangements and went off to Tokyo with absolutely no preparation. I was told to meet with our people at the U.S. embassy the afternoon before the meeting in order to get my instructions. The embassy officials told me I would have a dinner that night and that I was not to make any implied commitments of any kind. They said, "Do not nod your head from right to left, or nod your head up and down." Being adequately prepared by the diplomats to be a professional negotiator, I was taken to an elegant restaurant, where I sat on the floor between two geishas and tried to look straight ahead at the three Japanese diplomats who had many questions. Unfortunately, the geishas on both sides of me attracted most of my attention. They would offer me some sake or a piece of sushi, and I found myself looking to the right and left and responding to their kind offers by nodding my head up and down. I had totally failed to follow my diplomatic instructions. I knew I had to perform much better at the official meeting with the foreign minister.

I arrived at the appropriate official building and was escorted into a small ballroom with three rows of bleachers along one wall. These low steps were filled with photographers and photographic equipment, and I wondered what the Japanese Abe was planning. I found out in short order as he entered the room and shook my hand, and after a blistering onslaught of photographer's flashes, he abruptly walked out – without even a bow. Then the embassy folks immediately escorted me back to the airport for the long journey back to Washington. I never found out exactly what the press had to say about our meeting, but I did get a copy of a magazine that covered the event. I have the magazine filled with spectacular photos, but I never had it translated into English, so I will never know what they said I said.

The Cold War was reaching its apex, and the anti-SDI rhetoric was reaching new heights. Many who worked for the emerging SDI-related programs in Europe were being threatened when I was asked to go to another trip, this time to a well-advertised conference on SDI in Brussels. The climate of fear was increasing, and in my opinion, it had become a bit too risky, so I calmly refused to go. Abe reminded me that he was my superior officer and issued a direct order, but in my usual humor, I reminded Abe that I was a civilian and thus could not be ordered. Abe laughed, shook his head and said if that was the case, he would go by himself. He then said he had already arranged a security briefing to put my mind at ease, so I agreed to go.

A few days later, Air Force Colonel Stevens, a very clean-cut and rigid man, showed up at my office, saluted and stood at attention. "Dr. Yonas, I'm here to give you your security briefing for your trip."

"Don't waste your breath," I said. "I'm not going."

The colonel continued as if I had not spoken, "Doctor, you must follow these three instructions. First, when the terrorist walks down the aisle of the plane, he will hit passengers on the head with his rifle butt, so the doctor must choose a window seat." Aghast, I just stared open-mouthed at the colonel, thinking I was definitely not going on this trip. Without missing a beat, the colonel went on, "Doctor, this is your security briefing; listen carefully. Do you understand? When the terrorist comes into the room, jam your pocketknife between your legs." Now, I was really getting worried. What room? What pocket knife?

"When the terrorist comes into the conference room, jam your pocketknife between your legs, cut the seat cushion and hide your credentials," the colonel continued.

"Jam a knife between my legs into my credentials?" I asked. I was *definitely* not going.

The colonel just kept talking, "When you hear the explosion …"

I interjected, "You can't even stop them from blowing me up?"

The colonel continued talking. "When you hear the explosion, throw your body in the opposite direction since shrapnel cannot cause that much damage to your feet." I stared speechlessly at the colonel. "That was your security briefing, doctor. Did you understand?"

I said I totally understood and thanked the colonel, who saluted and marched out. Then I told Abe I was *definitely* not going. But Abe used the oldest weapon in his arsenal – guilt – and told me not to worry. He said he would go face the dangers alone and spread the goodwill of SDI and I could stay home and watch over the mundane details of the program. Of course, I agreed to go.

As I arrived at the airport in Brussels, I realized there was no escort waiting for me, so I took a regular taxi to the hotel and found Abe's bodyguard checking in. He had a duffle bag with two bulletproof vests. I smiled and asked him how they knew my size, and he said they were for him and Abe, not me. I just shrugged since I had become unaccustomed to the thought of any security preparations.

Later that evening, I had time to kill and went jogging, alone. I got lost and ended up jogging in my sweats in the red light district, business ladies sitting by large windows, waving at me – and me waving back. They might have been assassins in disguise, but they seemed harmless enough. Upon entering the first meeting the next morning, I noticed that (with the tight security) the metal detector at the entrance was not plugged in.

After the uneventful meeting with lots of questions about how to go about getting funding, I got a message that the Supreme Allied Commander of Europe, General Bernard Rogers, wanted to see me as soon as possible. I wondered what I had done at the conference to deserve this honor. I was told to show up at Mons, the base just outside Brussels, at an extremely early hour. As I arrived at a grand mansion, I was led to a massive room near a window overlooking a beautiful garden. Suddenly the entire building started to shake, and a loud "wopita, wopita, wopita" of a giant green helicopter took over. The massive craft landed down in the gardens right in front of me, the room's windows almost breaking from the air pressure. Out of the helicopter marched a movie star–looking, four-star general, resembling Paul Newman, with stunning gray hair and more stars than I had ever seen. General Rogers marched briskly through the beautiful garden and stormed through the French doors into the room. Without so much as a handshake, he proceeded to chew me up and down, reading me the riot act and claiming I was singlehandedly destabilizing the alliance. I respectfully thanked him for his advice and then proceeded to contradict his claims. He listened politely but did not seem impressed. Without any ceremony, he marched out and went back to the helicopter, which went "wopita, wopita, wopita" again on the way up, causing the windows to vibrate. When I returned to Washington, I found out that Rogers had gone public in an *Aviation Week* interview, accusing me of attending a conference and trying to destroy the essence of NATO.

During another trip, this time to Germany, to meet with potential "partners" and not subcontractors, I had a pleasant lunch with Karl-Heinz Beckhurst, one of the VPs at Siemens. Beckhurst made it clear to me that he did not like SDI, which he claimed was not going to protect Europe, and he refused to cooperate. But the mere meeting with me did not go unnoticed. He was murdered by a precisely timed, high-tech shaped charge that exploded in the rear seat of his limousine. The Red Guard left a note warning others not to get involved in the SDI. It was later reported that this assassination was carried out by the KGB. Shortly after that, some other European industry leaders were murdered, and my new German industry colleagues moved out of their houses into safe, undisclosed locations. There were also reported to be seven SDI-related terrorist kills and many other mysterious deaths of British scientists. Needless to say, I took the entire possibility of assassinations of our potential partners very seriously.

It was during this time that I learned my colleagues from my previous plasma physics research project were planning a conference. The meeting was scheduled to take place on one of the Greek islands. The plan was to discuss the state of the science on particle beams and lasers and also have a very good holiday. I was advised by people who should know about such things that there was very credible information regarding a plan, including rubber boats and machine guns, to kill all my buddies. Naturally, I had the meeting canceled, claiming some sort of snafu as an excuse. My friends thought I had put the kibosh on their conference because I was jealous since I could not go and I didn't want them to have fun without me. They never learned the truth.

I wondered, "If the bad guys wanted to kill me and my friends, and if the early deployment mafia wanted me gone, did that mean I was on the right track?" Of course my defense during these tense times was my sense of humor, although it was not always appreciated by most of my colleagues.

CHAPTER 28

The road to Reykjavik

IN 1986, THE SOVIET EFFORT that began in the 1950s to wage war in space was reaching its culmination as the Soviet Union's military industrial complex prepared to launch the world's biggest booster, which would be carrying a 100-ton demonstration of bits and pieces of the Soviet's first space-based laser. They called their space battle station Polyus and, with a laser payload, it was called Skif. This weapon could use rather small missiles or a giant laser to destroy U.S. satellites. The Soviets set 1987 as their target date for deployment. The clock was ticking, but their ace in the hole was the giant Energia booster, which was to be used for the first time in 1986 if they could only get the laser together in time. But the real clock was being set by the global politics that were spiraling out of control.

Gorbachev knew about Energia and Polyus, but it is not clear how much he knew. His closest technical advisor, Evgeny Velikhov, probably the leading scientist/engineer in the Soviet Union, must have told him of the limitations of the SDI technical program and the problems the Soviet labs were having with high-power lasers. After the failure of the Terra program and the untold billions of wasted rubles, Gorbachev was not anxious to proceed with another giant laser program, and yet the Polyus program was going ahead.

Influenced by his country's technological failures and the pessimism that followed the disastrous Chernobyl nuclear reactor explosion and resulting environmental catastrophe, Gorbachev was running scared. Then in August, the U.S. Senate voted to support the SDI, which made all the arguments against the Soviet program look inadequate. In less than two years on the job, Gorbachev had seen his country slipping into financial and technical oblivion, and there was Reagan with another initiative. What new technology surprise might the U.S. Star Wars program invent? So far, Gorbachev's attempts to halt the weaponization of space had backfired. According to news reports, when the U.S. demonstrated in June 1984 with the HOE the ability to "hit a missile with a missile," Gorbachev had met with British Prime Minister Margaret Thatcher to try to convince her to urge Reagan to put a stop to the SDI nonsense. When Thatcher told Reagan about Gorbachev's concerns, it gave Weinberger the ammunition he needed to convince Congress of the need for SDI funding. In 1986, the money began to flow.

Gorbachev had one last chance to put a stop to all of this, and that was through arms control negotiations. If he could convince the U.S. to back down with SDI, he could then convince his own military to back off. He did not want to engage in an arms race with the U.S. in his country's weak financial situation and, even though he was told by Velikhov that the SDI concept of defense against the Soviet missiles was foolish at best, he still feared that his own military could drag his country into another waste of money and resources they did not have. Gorbachev's goal was to use all his diplomatic skills in a one-on-one with Reagan at Reykavik, Iceland, in September 1986 to put a stop to all the space race competition. He knew that Polyus was about to launch, and he planned to do everything he could to stop the SDI and then turn around and stop his own program before things got out of control.

In October 1986, Gorbachev and Reagan met at Reykjavik for what could ultimately determine the survival of the planet in a growing proliferation of nuclear weapons. George Schultz, Reagan's secretary of state, helped the president prepare for the summit. I learned from Shultz's memoir that he really understood the limitations of our technology and was prepared to trade our nonexistent accomplishments for some meaningful agreement. In his instructions to Reagan in preparation for the Reykjavik summit, Shultz said, "We have to be willing to give up something in SDI. An agreement for massive reductions in strategic missiles can use SDI research and potential deployment of a strategic defense as a means to win Soviet compliance on continuing reductions. So we should give them the sleeves from our vest on SDI and make them think they got our overcoat."[92]

After almost three days of back-and-forth arguing about SDI and arms control, the leaders arrived at a dramatic point in the conversation that could have changed the world. The meeting was drawing to an exhausting and frustrating close, and it was either make or break when Gorbachev demanded dozens of times that the SDI "be confined to research and testing in the laboratories."[93] According to Ken Adelman, the U.S. chief of arms control, "Gorbachev's insistence that SDI be confined to laboratories…was repeated constantly … and then a stunning 20 times on Sunday afternoon. Mentioned every five minutes."[94] Gorbachev and Reagan had made it clear that they both badly wanted to abolish nuclear weapons, but they were stuck on the issue of testing outside of the laboratory.

During the last moments of the summit, Reagan stated, "It would be fine with me if we eliminate all nuclear weapons,"[95] and Gorbachev enthusiastically followed with, "Let's eliminate them. We can eliminate them." The world was on the edge of a fundamental change in military capabilities and nuclear weapon deployment. Then Reagan demanded to "continue research, testing, and development which is permitted by the treaty." Gorbachev objected. The Soviet leader said, "If development can go on outside the laboratory, and the system can go ahead in 10 years … It's laboratory or goodbye …10 years of research in the laboratories within the limits of the treaty ought to be enough for the president." [96] He was not against SDI. But the research had to be in the laboratories.

In hindsight, I believe Gorbachev was more worried about his own military industrial complex given the upcoming events at Baikanor. He was at the end of his patience and desperate to clinch the deal, but he knew he could not walk away and leave the door open for Baklanov to launch an ultimately dangerous and probably ruinous space arms race, and he must have been focused on the upcoming Soviet space launch event from which there would be no turning back. This world-changing agreement between the superpowers was hanging on one word: laboratory. I doubt if either of them knew the least bit about what that really meant, and for almost 30 years, I have been trying to understand what really happened. What were these world leaders really thinking? What else worried Gorbachev that none of us knew? I believe that while he argued with Reagan about future space weapons, what he feared most was what was about to take place in Kazakhstan. Meanwhile, no one in the U.S. was aware of the impending Soviet launch.

Baklanov, was convinced that both Gorbachev and Reagan lacked the technical understanding and capabilities to deal with these issues. "Reagan was completely illiterate when it came to talking about problems of a scientific and technical nature. He didn't understand anything he said about SDI ... a bluff and a myth," Baklanov wrote. "Gorbachev wanted to use the myth about the capabilities of SDI ... as a pretext for getting us to surrender ... Creating SDI systems in space would have required enormous and ultimately worthless expenditures."[97]

Baklanov was not only in charge of all military investments, but he became one of the leaders of the right wing of Soviet decision-makers that opposed Gorbachev's attempts at reform, and he was not alone in his opinion of Reagan. The nominal leader of the left wing, Alexander Yakovlev, Gorbachev's minister of propaganda, was in close agreement on the view of Reagan, saying, "In Reykjavik Reagan missed his chance to go down in history not as a clown, but as a statesman ... not intelligent enough and too limited in his freedom of choice ... systems have emerged against which there exist no effective counter-systems."[98]

I was determined to understand fully why we did not achieve an agreement and give up "the sleeves of our vest," so in 2014 I arranged a meeting with George Schultz, Reagan's secretary of state, who not only had understood the status of technology development and had a strategy for the negotiation but had stood by the president's side during the Reykjavik summit. When I asked Schultz why we did not trade, he made it clear that Gorbachev was convinced America had already developed successful SDI technology that could be extremely effective against their ballistic missiles, but the problem was that Reagan probably was so impressed by claims from his advisors and Gorbachev's negotiating emphasis that he was in no mood to give up the fictitious overcoat. I think Gorbachev protested too much, and maybe the emotions of the moment persuaded Reagan, but I wondered what was driving Gorbachev to walk away from a deal that would satisfy his dreams of world peace. I knew that the Soviet scientific community had provided substantial evidence to Gorbachev that the technical foundation of SDI was inadequate to justify the initiative, but I am not sure that scientific advice amounted to much when confronted with other more powerful forces.

The meeting with Shultz included Stanford physicist and arms control expert Sidney Drell. According to Drell and the details in the documentation he gave me, the information that Gorbachev had received from his political and military advisors before the summit had painted a frightening picture of American superiority.[99] Gorbachev was convinced that he had to stop, not the SDI as claimed by many, but to stop the arms race driven by his own military industrial complex. Before Gorbachev left for Reykjavik, he stated, "Our main goal now is to prevent another new stage in the arms race ... drawn into an arms race that is beyond our strength. We will lose because now for us that race is already at the limit of our possibilities."[100]

CHAPTER 29

Setting the record straight in Tutzing

I HAD COMPLETED MY TWO-YEAR assignment to the Pentagon, but I was nowhere near finished with my involvement in SDI. In December 1986, just a few months after I left Washington, I was invited to attend an SDI conference in Tutzing, Germany. I saw this meeting as an opportunity to close my assignment by "setting the record straight" and leaving a permanent record of what I understood at that time. Little did I know that many of my scientific colleagues attending the meeting were ready to take a particularly hostile stance against SDI. The participants were less interested in getting answers than in sharing their views of the past and future of the program. I usually took a less than serious approach to such public interactions, but this audience had little tolerance for my humor.

The meeting took place in a German castle near Munich in a building that looked a bit like a set from one of those World War II depictions of a U.S. command center. I could imagine General Patton standing at the front of the conference room giving orders. I later learned that General Eisenhower had in fact used the castle as his headquarters in command of the allied forces as the war drew to a close. The castle had been the centerpiece of the decisions involved in the destruction of Nazi Germany. At the time I did not know that we were involved in the soon-to-be destruction of the Soviet Union.

The castle had even a more notable history, as the meeting organizer Klaus Gottstein explained in the opening session. Gottstein surmised that in the eighth century, the inhabitants of the castle were worried about the coming of doomsday and the destruction of humanity in the 10th century. Gottstein explained to us that here, 1,200 years later, we were still "worrying about the possibility of doomsday," but he suggested optimistically that we were not just hoping, but pursuing SDI in order "to make peace more secure."[101] Despite Gottstein's optimism, many of the Americans present at the meeting were seeking an end to the SDI program, and they had arrived prepared to administer the program's final death blows.

The meeting included 33 carefully chosen participants who were encouraged to express their opinions openly concerning the murky social, political, economic and even technical issues of SDI. Most of the participants were German scholars and defense professionals, but the attendance included a handful of carefully chosen scientists and academics. I knew all the Americans since I had been engaged often in debates and discussions with them and knew they would be opposed to my opinions.

I saw this meeting as an opportunity to tell my story and deal with the most difficult of all audiences I had ever faced. I gave the opening presentation with a detailed and fairly complete view of the myriad technical issues that had to be resolved. Many of these had been identified in the summer of 1983 study chaired by James Fletcher. We had only just started on the program in the spring of 1984, and in less than two years of real organized activity, there had been little real resolution of any of the key issues. I had no illusion that my audience would tolerate my lack of reported results. As expected, they were not at all forgiving.

The atmosphere was initially very informal and friendly with suggested audience participation, but it did not take more than a few minutes into my talk for Richard Garwin, probably the most knowledgeable strategic offense/defense expert in the world, to interrupt me and say, "Enough about questions, but what are the answers?"[102] He knew from my opening remarks that I viewed SDI as a long-term, broadly based research program, but he would not accept that that we really didn't know the answers already, due to the many administration claims. I argued that we would have to spend several tens of billions of dollars and at least 10 years to answer the important questions, but Garwin disagreed.

Garwin's interruption was just the first salvo of what would be a determined onslaught from the American defense experts. I felt the Germans wanted to be more open and even sympathetic to my views, maybe out of courtesy to a guest, but the Americans showed little kindness to the person they treated as a rank amateur or even a slightly unprepared scoundrel. I was faced with blistering comments from some of the most highly respected arms control experts.

Robert Cooper, a recent head of DARPA who had been engaged for many years in strategic defense technology, did not hide his contempt for what he described as the low technical capabilities of the Department of Defense. Cooper even implied that I was an example of the DOD's lack of knowledge. He made it clear that there was nothing new in missile defense science and technology to back up the Reagan vision to "protect all of the people all of the time." Cooper opened the gates for Jack Ruina, another former head of DARPA, who then went on the attack. "Gerry won't like this, but Star Wars has seen its last good days, and I would not buy stock in the SDI program now. We don't always believe in what presidents have said in the past, and after all, this is political rhetoric."[103]

Garwin listened for a while, and I began to think he was going to drop the personal attacks and defend me when he said, "Of course SDI has some good things going." I sighed with relief, but then he went on to say, "Even the Mafia has some good things they do ... and when the SDI is put into receivership under the bankruptcy lawyer, there will be an official appointed to look at the good programs."[104] I wondered if he might want to be that official.

George Rathjens, a distinguished MIT professor and a former deputy head of DARPA, added, "SDI is going nowhere, and we should not support it any longer ... the responsible thing is to kill it as gracefully as we can."[105]

I argued for a continuation of the program that would lead to an arms control agreement that allowed for a continuation of a treaty-compliant technology program. I suggested that we consider other forms of retaliation that would be less time-urgent and thus avoid instabilities in a rapidly evolving crisis. I advised that we move away from missiles delivering multiple warheads and toward an agreement that would be based on a public understanding of the problems of continuing the present approach. I concluded that rather than continue the meaningless discussion of a perfect defense, we focus on a more stable form of deterrence. I argued, "The public will accept the impossibility of eternal life, but would be willing to invest in research to prolong life."

My point of view was one of strengthening deterrence, rather than getting rid of it and jointly managing a transition with the Soviets to greater reliance on defense and reduced investments in strategic missiles. I suggested, "Societies will change if we can survive until sometime in the next century, where we may look upon this reliance on a vast quantity of nuclear weapons as being a temporary chapter in the history of mankind."[106] In the end, the conference participants only wanted to talk about the futility of the technology. I didn't disagree with the futility of the technology, but technology is only useful within the political context. Like Freeman

Dyson, I was seeking a live-and-let-live, win-win solution to ending the arms race. But my American colleagues were not listening.[107]

The conference in Tutzing ended with the SDI detractors muttering that the U.S. should stop wasting money and the Germans still wondering how it was possible that the Soviets could be so worried about such a "worthless" program." Little did we know at the time that the Soviets had taken the U.S. SDI program very seriously and their investment in the arms race had helped drive the country to economic ruin. [108]

CHAPTER 30

The Polyus launch

SHORTLY AFTER THE REYKJAVIK SUMMIT, Baklanov brought his team of scientists and engineers to the Baikanor Cosmodrome, the Soviets' space launch facility in Kazakhstan. They were there not just to view an historic rocket launch, but to persuade Gorbachev not to stop it. When they arrived, there ready for launch was the world's largest booster, Energia. It was carrying the 100-ton Polyus space-based laser demonstration and, after of months of preparation and then a delay for the arrival of dignitaries, the Soviets were ready for the Energia's historic first launch. Gorbachev and members of the Politburo arrived, and the Soviet leader immediately declared to the already exhausted and frustrated launch crew that there would be no launch. He ordered them, in classic bureaucratic style, to do more analysis and write comprehensive reports. Then to make matters worse, he preached to them about the evils of space-based weapons.[109]

Baklanov was well aware of Gorbachev's attitudes, and he was prepared for this move. That evening he and his team gave Gorbachev a comprehensive briefing on the subject of rockets, space and Energia. He undoubtedly emphasized the glory of Energia and Soviet space technology, saying, "We created close to 85 new materials of a higher caliber than anything else in industry and engineering ... we introduced something on the order of six hundred innovations."[110] I can imagine how a technically educated person might be persuaded by such argument, but none of this should have been convincing to a social/economics expert or a diplomat who did not care about any technical innovations. After all of Gorbachev's Reykjavik furor over the need to stop development and deployment of any weapons in space, how could he relent? But surprisingly, Gorbachev gave the go-ahead for the launch the next day.

I was confused at first that Baklanov had convinced Gorbachev to change his mind and abruptly reverse his conviction to stop the launch of Energia – until I learned a great deal more about Gorbachev's behavior from the subsequently very revealing book and interview with V. Boldin, one of Gorbachev's closest associates for 10 years and his chief of staff. Undoubtedly, Boldin's comments, including his book, which he wrote behind bars after he participated in the failed coup to overthrow Gorbachev, were driven by intense dislike or worse, but I don't think his words can be entirely ignored. Boldin said, "Gorbachev is a weak-willed proud and unstable person ... changed his position and was too pliant ... never wanted to say anything firm to anyone, to make any reproach or exhibit the toughness required by a particular situation ...

everything could be turned on its head on a daily or hourly basis." In his memoir, he described how Gorbachev, "for whom maneuvering had become a habit, was really taking two steps forward, three to the side, and one backward, and everyone found such conduct disconcerting ... he had everybody confused."[111]

I had heard a similar description of Gorbachev's behavior from a credible Russian scientist who was close to the negotiations but chose not to be identified. In another blistering comment Boldin said, "Gorbachev is a coward by nature."[112] Strong words from a bitter man who turned against his boss of 10 years, but similar comments were made by Baklanov, who described Gorbachev as a manipulator and schemer, calling "SDI a bluff and a myth. Call it what you will. Perhaps Gorbachev wanted to use the myth about the capabilities of SDI for certain purposes ..., such as foil for his later actions ... I came to think that SDI was really used as a pretext for getting us to surrender."[113] Very sour and even bitter grapes, but not contradicted by the comprehensive and well-documented historical analysis of Vladimir Zubok, who wrote, in my opinion, the most convincing description of the period. He quoted William Odom, who was the director of the National Security Agency: "Gorbachev was an inveterate schemer, loquacious obfuscator, unable to anticipate the likely consequences of policy."[114] Zubok also quoted A. Dobrynin, Soviet ambassador to the U.S., writing that Gorbachev "had the emotional makeup of a gambler ... was visible even in 1986 at the Reykjavik summit."[115]

So I surmise that Gorbachev could easily have changed his mind on the spot, depending on the input from the last persuasive person. He had the ad hoc optimism of the moment to hope things would work out, and he lacked the inhibitions to be consistent. But, what were the relevant traits of his negotiating partner, and why did he not compromise? Much has been written about Reagan, but the man who worked closely with him for 30 years was Michael Deaver, who became the deputy chief of staff in the White House and was close to both Reagan and his wife. In his memoir, he captured an insight into Reagan's character that seemed to me to be at play at Reykjavik.

According to Frances Fitzgerald in her comprehensive analysis of Reagan, Deaver believed that Reagan "was superstitious ... consulted his horoscopes in the newspapers ... believed in luck and fate and events beyond human control... world of theater ... everything on a given evening is beyond their control."[116] Maybe Reagan had no choice but to follow his imagined script. He was acting out his part as the savior of his country, and he had to fulfill his obligation to the American people to deliver the ultimate defense, like an umbrella in a rainstorm, which would stop incoming nuclear warheads. Unfortunately, the reality was that even in the most optimistic outcome, the umbrella would be useless in the torrential downfall of the hurricane of an all-out nuclear missile attack.

Maybe Reagan's belief in fate and luck was not misplaced since the momentous event of the start of a real Star

Wars did not occur. Circumstance, incompetence or luck played the key role in the outcome of the Energia launch. Ironically, the launch succeeded perfectly, but a software glitch caused the orbital deployment to fail.[117] Polyus went unceremoniously into the Pacific and was never seen again, and Gorbachev was saved from the inevitable U.S. reaction and Soviet counteraction.

CHAPTER 31

The beginning of the end

By the end of 1987, the entire Soviet space weapons business was finished, and in only two more years, so was the entire Soviet Union. The Terra program and the Polyus program came to nothing. Much could have been different had the Soviets not been deluded by the fiction of high-power lasers. Had they not been seduced by the false glamour of high-power lasers, had the Polyus gone into orbit carrying a bundle of small rockets, and had Gorbachev not been fooled by the science fiction of SDI and the misguided determination of Ronald Reagan, then the real Star Wars would have been off and running.

As it was, Gorbachev was determined to avoid a space weapon competition with the U.S. He could have had a strong hand to negotiate what he wanted on his terms, but he managed to play with a weak hand against an even weaker hand, and both sides walked away from what could have been the beginning of the end of the nuclear arms race. The phony glamour and delusion of high-power lasers prevented both sides from finding a way to deliver on Reagan's dream to do away with all nuclear weapons, although they did manage to walk away from the creation of real war in space, and they did move on to real arms control. In June 1987, Reagan gave his prophetic Berlin speech, "Mr. Gorbachev, tear down this wall." In December 1987, the two sides agreed to get rid of all intermediate-range nuclear weapons, and this seemed to end the Cold War.

After Reykjavik, Gorbachev was given an opportunity to fire most of the top brass in retribution for an incident in which a German pilot flew a small private airplane from Finland into Moscow near the Kremlin. Subsequently, exiled H-bomb physicist Sakharov returned to Moscow and told Gorbachev not to worry about SDI and to get on with other agreements. Gorbachev did just that. He proceeded with the economic revolution that was failing miserably in spite of, or possibly due to, many of his misguided efforts. The problem was that he wanted to keep communism and also create a competitive entrepreneurial economy. This proved to be a fundamental contradiction that could not be solved by Gorbachev's rhetoric.

Meanwhile, while the Soviet economy continued to sink, Reagan, with his Ollie North/Admiral Poindexter-run Iran/Contra scandal, and Gorbachev, with his sober but failed economy, agreed to several arms control agreements that just further infuriated the Soviet military industrial complex. By 1991, Baklanov and Valery Boldin, Gorbachev's chief of staff for 10 years, along with six others, including the chief of the military, staged a coup and deposed Gorbachev.

The coup plotters delivered the blow with the help of the KGB, by invading Gorbachev's summer house in the Crimea. Baklanov described the encounter in an interview in 1999: "Gorbachev was scared … his anxiety was visible … shaking violently … a dull man thinking in a dull way about himself rather than the matter at hand." As it turned out, the plotters were not very good at planning, and the coup was wiped out in a few days. Boris Yeltsin took over, and Gorbachev's economic revolution, along with the Soviet Union, was gone.

The historian Pavel Podvig, after an extensive study of the comprehensive Soviet archives that appeared after the Cold War, argued convincingly that the vigorous attempt by Gorbachev to curtail SDI was really a fear of unleashing the powerful Soviet military industry complex on an uncontrolled Star Wars spending spree.[118] But if that is true, I asked myself, then what could Gorbachev have been thinking at the closing moments of the Reykjavik meeting that prevented him from realizing his goals to end the arms race, abolish nuclear weapons and keep SDI within the bounds of the existing agreements? Maybe there was something else; maybe he had been given information before the summit that had a profound impact on his attitude. Maybe he had some scheme up his nonexistent sleeves?

I went back and spent hours searching through Gorbachev's obtuse and confusing memoir that seemed to me to be a self-serving account, admitting not a single mistake.

I was trying to find some previously hidden Gorbachev perspective on what happened at Reykjavik. I found that he barely mentioned strategic defense, or Reagan, although he certainly deeply believed in abolishing nuclear weapons, and he said in his proposal in January 1986 to do away with all nuclear weapons was "not utopian after all ... this noble and salutary goal is reachable, given the goodwill of all members of the international community."[119] He never indicated that the SDI had any impact on ending the Cold War but instead wrote, "The Cold War was brought to an end thanks to Perestroika and the new thinking." He also made it clear that the "totalitarian system had run its course morally and politically, and a prolonged and potentially deadly period in world history, in which the human race had lived under the constant threat of nuclear disaster, had come to an end."[120]

I think Gorbachev provided the final word on the question of the explanation of the end of the Cold War, since he took primary credit for making it happen, but he missed his chance at Reykjavik to start down the path to remove the nuclear danger. Zubok described the collapse as a failure of will to save "the empire they did not believe in, and for the empire they did not profit from. Instead of fighting back, the Soviet socialist empire, perhaps the strangest empire in modern history, committed suicide."[121] Maybe this self-inflicted wound was not that dramatic but was just a compounding of very bad management. Boldin wrote, "By 1987, virtually the entire membership of the Poliburo had been changed, only to undergo another overhaul in 1990 ... utterly incapable of deciding or uniting anything at all – a sure sign that the collapse of the organs of government and of the entire party was imminent."[122]

The concept that one of the largest and most powerful countries in the world committed suicide because of its moral decay and mismanaged political institutions, rather than economic collapse or even a scientific and technology competition, as claimed by many, is profound. It is a cautionary tale about protecting and unifying the national social and political fabric as well as its military-based national security that needs to be a warning to us today.

CHAPTER 32

After SDI

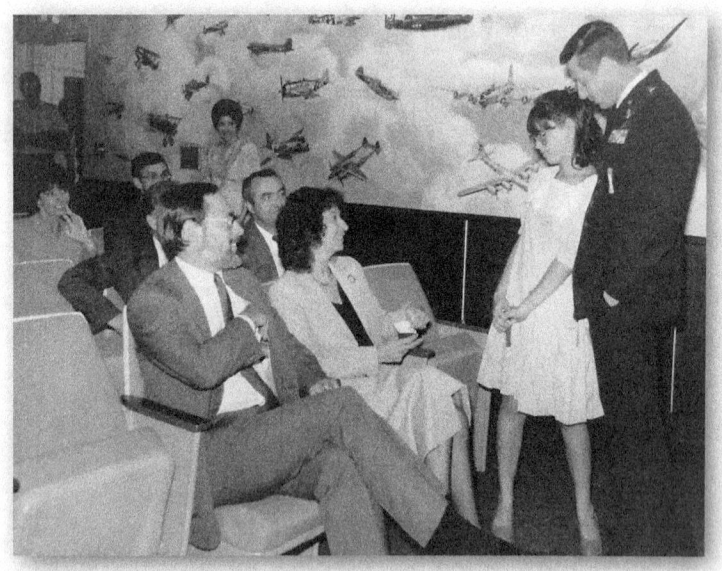

Abe presents Jane and Jodi Yonas with gifts at a farewell ceremony as Yonas' time with the SDI draws to a close.

AFTER I LEFT THE SDI program in 1986, I wanted to pursue entirely new subjects in the private sector and even escape from the subjects related to missile defense that I had been pursuing for decades. I joined the Titan Corporation primarily because it was committed to growing through acquisition of small, high-tech companies. I knew nothing about the acquisition business, and that became my assignment. My new job came with some old problems. As usual, there were questions about legal issues of conflicts of interest following accusations that I had joined Titan to shovel SDI contracts to them under the table. The San Diego paper had investigated claims that l was sending contracts to my former employer, Sandia Labs, and they were ready to go to print when I informed them that I no longer worked at Sandia. Undaunted, they decided to pursue my Titan activities but found I had shifted my interests to subjects unrelated to the SDI, such as accelerators for physics applications. I also was very interested in commercial applications of technology, but I soon found out that my business acumen was not terribly impressive.

My escape from SDI was not entirely successful. After Tutzing, I felt that I had done my duty, but in early 1987 I became involved in many studies to evaluate what had been done and what to do next. The White House Science Council (WHSC) created a six-month study chaired by Sol Buschbaum from AT&T, a friend who had influenced much of my career. The panel included Edward Teller and Harold Agnew, who were knowledgeable about everything that had happened over the past few years. I was surprised that the panel was primarily interested in near-term deployment, but

Gerold Yonas, Ph.D.

Upon leaving the SDI program, Yonas receives these framed medals signed by Caspar Weinberger. The award reads:

To Gerold Yonas for exceptionally meritorious service to the country by significant contributions to the Nation's Strategic Defense Initiative as the program's first Chief Scientist and Acting Deputy Director. A renowned scientist, specializing in high energy particle beams and pulse power technology, Dr. Yonas was called upon to participate in the Defense Technology Study Team. Tasked by the President of the United States to make nuclear weapons "impotent and obsolete' the members of the team reviewed al existing technologies, analyzed the threat, and outlined a feasible course of action the nation could follow to examine the feasibility of true strategic defense. This study, subsequently known as the Fletcher Study, became the basis for the national strategy on Strategic Defense and the foundation of the new Strategic Defense Initiative Program. Dr. Yonas' untiring work and dedication has helped to establish the Strategic Defense Initiative Organization on a firm scientific foundation and a strong technology management base. I take pleasure in presenting Gerold Yonas the Secretary of Defense Medal for Outstanding Public Service. Caspar Weinberger, Secretary of Defense, August, 1986 they had no interest in arms control, which I thought should be the real focus of future SDI work in order to deal with the many unanswered technical issues. The panelists focused on ground-based interceptors and had no interest in long-term research and directed energy weapons. It seemed to me that they were turning the clock back five years.

The panelists were also influenced by a recent complete review by the Defense Science Board (DSB) that had concluded that the program should first prove we could find real targets and manage the battle before we got too serious about intercepting targets. The DSB was critical of the SDI management because of its lack of progress in system integration, and they recommended the creation of the SDI Institute as soon as possible. The panelists were particularly discouraging about any sort of early deployment except for silo defense. They also managed to take a shot at the SDI's focus on expensive sales-orienteddd demonstration projects that they thought were technically unjustified.

Although the WHSC reaction to the DSB report was not unanimous, there was a consensus to focus on deployment of a system with limited capabilities and to deal with real engineering issues of intercepting targets with homing missiles. I was surprised that the panelists recommended a treaty-compliant, ground-launched interceptor possibly deployed at Grand Forks, North Dakota, to take advantage of the existing infrastructure. This was turning the clock back decades and totally ignored any possibility of ever reducing the threat through negotiations with the Soviet Union, which was in an early stage of collapse. The panelists stated that in the far future, some sort of space-based interceptor, probably small missiles, would be useful and, as usual in such committees, they supported an aggressive R&D program. They did emphasize that the investment strategy should have high priority for finding, identifying and separating real targets from decoys and tracking the targets by interceptors.

The support for early deployment continued with another DSB study in early 1988. This study's first priority was sensors and information management, but the participants got serious about deployment. The study supported deployment of 100 ground-based interceptors and urged the Pentagon to get serious about defending Washington from short-range, submarine-launched ballistic missiles, cruise missiles and stealth aircraft. This study's concerns were well placed and had been ignored by the SDIO. We certainly did not have a clue about any sort of defense against these threats. The cost of seriously dealing with such defense issues was beyond my wildest comprehension.

Neither the DSB nor the WHSC gave any consideration to what was politically, economically and psychologically going on in the Soviet Union, nor was there any interest in the president's total commitment to abolish all nuclear weapons. None of us seriously believed that Gorbachev would shift his defense investment to civilian needs, and none of us could even consider the most remote possibility of the end of the Soviet Union. By the end of 1988, the Soviet Union was on its way out, the Baklanov space weapons initiative was dead, our support for directed-energy weapons was on its way out the door, and nobody was interested in abolishing all nuclear weapons. The SDI budget was imploding about as fast as the Soviet Union.

Death Rays and Delusions

To Gerry Yonas
With best wishes,

Ronald Reagan

A signed photo from Ronald Reagan is another parting gift.

CHAPTER 33

A lesson and hope for the future

AND SO ENDS MY TRAGIC tale of the SDI program that led to a summit of two men in Reykjavik who desperately wanted to abolish nuclear weapons but ultimately failed to reach an agreement that could have changed the world. MacFarlane said the SDI was a sting; Woolsey said it was the rhetoric of speech writing; Keyworth called the directed nuclear weapons "unadulterated lies." Mikhailov called directed nuclear weapons an "evil jinn." Schlesinger called SDI "a bargaining chip" and Baklanov called it a "hoax." Today I think that although all these perceptions were somewhat real, they don't tell the whole story.

Personally, I now consider the SDI and the events that unfolded at Reykjavik to be "a lesson" about wicked problems that have no straightforward linear solutions but are still amenable to progress nevertheless. The key to such progress is patience and persistence. In the nearly 30 years since I served as the first chief scientist and acting deputy director for the Strategic Defense Initiative, I have learned that what often seems initially to be just one more engineering problem is actually a far more complex combination of social, economic, political and technical challenges. These types of complex problems were described in the systems engineering literature as "wicked," and the solution path was far from the straightforward linear approach I had been following for much of my career.[123] I have learned that there are no useful opportunities to disassemble the problems into their components, solve the components and then reassemble the pieces into a solution. The only useful approach is to deal with each problem as an integrated "mess" and then devote much of my effort into working on the entire mess.[124] I began to think of myself as a wicked engineer who learned to deal with what I naively thought of at first as unique and brilliant insights, which became the discouragement of dead ends and false understandings, followed, if I persisted, by eventual glimpses of reality through a fog of confusion. A good approach to solving wicked problems is summed up in a quote often attributed to Churchill: "Success is not final, failure is not fatal; it is the courage to continue that counts."

This story is a complex lesson about not just technology as I had first thought, but about politics, psychology, leadership, competition and control. Instead of SDI ending the Cold War, which was on its final path on its own, it had stood in the way of an agreement to abolish nuclear weapons. Had we learned this lesson back in the 1980s, perhaps we could have taken steps to begin to abolish nuclear weapons and create a relationship of mutual understanding and trust that escapes us even now. The key was then – and still is – the art of empathetic communication between two very different cultures. Peace might have been possible between two men who were both dreamers and ideologues. If Gorbachev were such a gambler, he should have taken a roll of the dice, and Reagan should have gone along with his new friend. Reagan asked Gorbachev "to change his mind as a favor to him, so that hopefully they could go on and bring peace to the world,"[125] but Reagan could have just as easily gone along with global abolition of nuclear weapons as the prize of an SDI agreement. Instead he was stuck on his SDI vision that stood in the way of his real goal. Reagan said, "I had pledged to the American people that I would not trade away SDI – there was no way I could tell our people their government would not protect them against nuclear destruction."[126]

The immediate reaction to even the hint of nuclear abolition was negative and vicious on all sides of the political and military spectrum. Senator Sam Nunn, one of the most respected arms control experts in the Senate, said, "It would have been the most painfully embarrassing example of American ineptitude in this century."[127] According to Fitzgerald, "Thatcher descended like a thunder-cloud upon Washington … she consulted with other NATO leaders … and all agreed that that Reykjavik had been a folly."[128] If Thatcher and Nunn were skeptical of an agreement, should any of us really hold out hope for a different relationship in light of the history of the Soviet Union?

Reagan grew to trust Gorbachev, but he was also fond of saying about any arms control agreements with the Soviets, "trust but verify." Many experienced in the history of the Soviet Union offered little reason for trust. No person was more pessimistic about a new relationship than Alexander Yakovlev, Gorbachev's chief of propaganda, who is credited with being the originator of much of Gorbachev's ideological theory of perestroika and wrote a comprehensive history of the evils of the Soviet Union. He claimed that the Soviet system was "responsible for the deaths of at least 60 million Soviet citizens … it is still enslaved by an official imperial ideology, the essence of which is that the state is everything, and the individual nothing."[129] Yakovlev ended his scathing recounting of the bloody history of Bolshevism with his statement: "Only when it has shaken free of Bolshevism can Russia hope to be healed."[130] It is not clear that the "healing" of the Russian psyche is complete even today, and it was certainly less so immediately after Reykjavik, when decision-makers throughout the world were stunned by what had almost happened.

The fallout from the Reykjavik non-agreement was similar to that from Reagan's 1983 announcement of the Star Wars initiative to "make nuclear weapons impotent and obsolete." Both Gorbachev and Reagan shared a vision, but neither had a realistic idea how to implement it, and it caught everybody off guard. The fight between the left and the right in both societies would have been furious, but in my opinion, the route to an agreement was emerging and should have been seriously pursued.

The academics would have had to walk back their predictions in 1986 that "Gerry won't like this ... SDI has seen its last good days."[131] They would have had to stop attacking the program's science and get on with solving the technical challenges in a jointly managed program, probably with the U.S. paying the bill, and focused on the computer problem of battle management. The military and industry leaders who were lusting after new programs would have found nonmilitary challenges to apply their facilities and capabilities. The diplomats would all have been given satisfaction in continuing negotiations, and they would have found many ways to argue endlessly about the nuances of verification of agreements.

It might have taken years, but maybe we could have aided the "Russian psyche healing" process and avoided the resurgence of the Russian militaristic strategies and investments. Or possibly, changing the Russian culture is a totally foolish concept? Maybe it was just an unrealistic dream after all, as now the nuclear arms race is heating up again in the United States, Russia, Korea and possibly Iran.

CHAPTER 34

What next?

AFTER THE COLLAPSE OF THE Soviet Union in 1991, a solid-state laser scientist in Moscow invented an efficient and rugged diode-pumped laser, and following a few moves and private investments, IPG Photonics, now based in Massachusetts, developed the most powerful commercial fiber lasers in the world. Recently Lockheed Martin claimed that they have developed an efficient, rugged, 30,000-watt fiber laser weapon that might someday be scaled to the megawatt (MW) class. The laser company, IPG Photonics, is now worth almost $8 billion and their products are used all over the world in auto manufacturing. One day these lasers may also be used for missile defense. Probably the biggest tech problem will be keeping the laser cool, because even with the most efficient devices in the MW class, operation has to remove MW levels of waste heat. This seems to be solvable, and I believe laser weapons will finally emerge.

The Soviets have improved their ICBM capabilities, as evidenced by their 2016 test firing of the 10-warhead, submarine-launched Bulava missile, which has advanced countermeasures against ballistic missile defense. This capability is more worrisome when we take into consideration that in 2014 Gorbachev warned that tensions between the U.S. and Russia over Ukraine have put the world "on the brink of a new Cold War."[132] In addition, we now are witnessing a continuing threat from nuclear weapons and long-range missile proliferation that appears to be growing moment by moment. The lost opportunity in Reykjavik is becoming a distant memory, but I still can dream, as I did in 1981, that Dyson was right about an optimistic conclusion: "If we decide on moral grounds that we choose a defense-dominated world as our long-range objective, the political and technical means for reaching a defense-dominated world will sooner or later be found, whether the means are treaties and doctrines or radars and lasers."[133]

As I now reflect on global relations, I wonder if America can ever become more trusted, thoughtful, rational, empathetic and well-liked in the world. Can the country develop a willingness to 'live and let live' that will allow us to avoid the next military challenge coming from the proliferation of nuclear weapons? Or will an attempt to learn to get along simply encourage the bad guys of the world to try to take advantage of us? Meanwhile, with increasing proliferation of long-range missiles and nuclear weapons, maybe technical breakthroughs and changing political thinking will create new opportunities for a credible missile defense. Scientific developments now appear to be making a true Star Wars program possible – but whether such a program would help or hinder world safety and stability remains unknown.

More than 30 years after Reagan described his vision of rendering nuclear weapons impotent and obsolete, more countries are gaining nuclear capabilities and political instability appears to be growing. Nuclear weapons may be falling into the hands of terrorist extremists. Deterrence is becoming increasingly complicated as the number of adversaries multiplies. As we enter this second nuclear age, we must face the fact that conflict between small nations with growing nuclear arsenals could have devastating consequences across the world. I believe science and technology can play a part in preventing nuclear holocaust, but my time working with the SDI has shown me that people, politics and perceptions will have an even more important role.

With all these unknowns, my number one piece of advice to both politicians and scientists is, "Don't lose your sense of humor." As I look back at the years I spent working on the SDI, I realize the important role humor has played in helping me cope with the nonlinear chaos and numerous uncertainties that have marked my career. Whether I was accepting the title of the Father of the Pluton Beam, fending off ludicrous accusations of hiding alien spacecraft or playing jokes on Abe, I made sure never to take myself too seriously. My sense of humor allowed me to deal with the daily chaos, confusion and contradictions in my SDI assignment and the deadly serious implications of much of my life's work.

In the end, my SDI experience is not a story about technological developments, engineering or science. Instead, it is a tale about the importance of perceptions, about scientific advisors struggling to inform political leaders and about politicians carrying out complex strategies based on technological fiction. It is a story of economic collapse, declining faith in science, social instability, delusions, deceptions and the pursuit of an impossible death star. Finally it is a story about hard work, laughter and cross-continental friendships and opportunities and experiences I will never forget. Much like any popular science fiction novel or movie, a sequel to my story is forthcoming … but for now, we have to wait for its release.

ACKNOWLEDGEMENTS

I WANT TO ACKNOWLEDGE THAT the development of this book began with extensive interviews by Nigel Hey for his book, *The Star Wars Enigma*. I was also prompted to write by Adrian Rudamin who encouraged me to collect vignettes for a possible TV series or movie, which never developed, but led to this book.

I apologize to my many friends, particularly Tom Blau, who suffered through early drafts of this book and I thank them for their support and patience.

I want to thank Robert Hummel for publishing excerpts from this book in the *Science, Technology, Engineering and Policy Studies (STEPS)* magazine of the Potomac Institute for Policy Studies.

I will always be grateful to Jim Abrahamson (Abe) who took a crazy gamble when he asked me to help him in 1984 and suffered through my sometimes insubordinate pranks.

Most importantly, I want to thank my wife Jane--my partner, girlfriend and soulmate--for more than 60 years of love, support and much needed criticism.

ABOUT THE AUTHORS

Dr. Gerold Yonas

DR. GEROLD YONAS HAS CONSULTED for numerous national security organizations including the Defense Science Board, DARPA, the Air Force, the Army, the U.S. Department of Energy and the Senate Select Committee on Intelligence. He is a Fellow of the American Physical Society and a Fellow of the American Institute of Aeronautics and Astronautics and

Gerold Yonas, Ph.D.

has received many honors, including the U.S. Air Force Medal for Meritorious Civilian Service and the Secretary of Defense Medal for Outstanding Public Service. Yonas has published extensively in the fields of intense particle beams, inertial confinement fusion, strategic defense technologies, technology transfer and "wicked engineering." He is a senior fellow and member of the board of regents for the Potomac Institute for Policy Studies.

After serving as the acting deputy director and chief scientist during the implementation of the Strategic Defense Initiative, Yonas went to work for Titan Corporation in San Diego, where he managed a group of small research companies. Three years later, he returned to Sandia National Laboratories to lead the pulsed power fusion program and several weapon related programs in the role of vice president of Systems, Science and Technology.

At Sandia, Yonas went on to create the Advanced Concepts Group and explore new opportunities including brain research. Following his retirement from Sandia in 2009, he joined the Mind Research Network as the director of neurosystems engineering where he explored the link between neuroscience and systems engineering. He also developed a graduate course in this field and taught as an adjunct professor in the Department of Electrical and Computer Engineering at the University of New Mexico.

Yonas holds a Ph.D. in engineering science and physics from the California Institute of Technology and a bachelor's in engineering physics from Cornell University, where he also received a varsity lightweight crew letter. He is married to his high school sweetheart, Jane, and is the father of two daughters, Jill and Jodi, and the grandfather of five children--Libby, Jenna, Jonathan, Emily and Ben. Yonas makes his home in Albuquerque, New Mexico, with Jane and the family pet, Peter the Dog.

Jill (Yonas) Gibson

Jill Gibson is the associate dean of liberal arts, the coordinator of the Matney Mass Media Program and the co-coordinator of the honors and scholars programs at Amarillo College. She has spent the past 20 years in higher education both as an administrator and faculty member. Gibson has also worked as a television anchor, reporter and producer. She holds a master's degree in journalism from Northwestern University and a bachelor's in English and drama from Stanford University. She enjoyed having the chance to help her father, Gerold Yonas, with the writing of this book.

ENDNOTES

---

1 Robert S. McNamara, "Mutual Deterrence" (speech, September 18, 1967), http://astro.temple.edu/~rimmerma/mutual_deterrence.htm

2 "JFK on Nuclear Weapons and Non Proliferation," Carnegie Endowment for International Peace, http://carnegieendowment.org/2003/11/17/jfk-on-nuclear-weapons-and-non-proliferation-pub-14652.

3 "Nuclear Test Ban Treaty," John F. Kennedy Presidential Library and Museum, http://www.jfklibrary.org/JFK/JFK-in-History/Nuclear-Test-Ban-Treaty.aspx?p=2.

4 Freeman Dyson, "Bombs and Poetry," The Tanner Lectures on Human Values, Brasenose College, Oxford University. May 5, 12, 19, 1982, http://tannerlectures.utah.edu/_documents/a-to-z/d/Dyson83.pdf.

5 Ibid.

6   John Kenneth White, "Nationalist Republicans and the Cold War," excerpted from *Still Seeing Red: How the Cold War Shapes the New American Politics,* Boulder, CO: Westview Press, 1997, pp. 138-142, http://www.ropercenter.uconn.edu/public-perspective/ppscan/86/86050.pdf.

7   Lou Cannon, "Actor, governor, president, icon," *Washington Post,* June 6, 2004, http://www.washingtonpost.com/wp-dyn/articles/A18329-2004Jun5.html.

8   "The Challenge of Peace: God's Promise and Our Response, A Pastoral Letter on War and Peace by the National Conference of Catholic Bishops," May 3, 1983, http://www.usccb.org/upload/challenge-peace-gods-promise-our-response-1983.pdf.

9   Murrey Marder, "Analysts say Andropov status beyond repair," *Washington Post,* Nov. 12, 1983, http://www.washingtonpost.com/archive/politics/1983/11/12/analysts-say-andropov-status-beyond-repair/ad782b1b-a339-4fe2-bb82-23371ba4755b/.

10 H.G. Wells, *The War of the Worlds* (London: Chapman and Hall, 1898).

11 Alexi Tolstoi, *The Garin Death Ray* (Moscow: Foreign Languages Publishing House, 1955).

12 "Buck Rogers Ray Gun Toy, 1934," *Smithsonian Snapshot*, http://newsdesk.si.edu/snapshot/buck-rogers-ray-gun-toy.

13 Robert Wise, *Robert Wise on His Films*, Silman-James Press; 1st edition. August 1, 2006, 107.

14 "Tesla," *Twenty First Century Books*, http://www.tfcbooks.com/teslafaq/q&a_011.htm.

15 "Murder in the Air," Turner Classic Movies website, http://www.tcm.com/this-month/article/218460|218498/Murder-in-the-Air.html.

16 Benjamin Wakefield, "A History of the Death Ray," *Strange Horizons*, Nov. 9, 2009, http://www.strangehorizons.com/2009/20091109/wakefield-a.shtml.

17 "Successes and failures of the 1970's," August 22, 2006, https://sputniknews.com/voiceofrussia/2006/08/23/102485.html.

18 P.V. Zarubin, *History of High Energy Lasers and Laser Based Systems*, 2004.

19 Ibid.

20 Ibid.

21 Reprints from *Aviation Week and Space Technology*, May 1977-Nov. 1978, McGraw Hill Inc.

22 H.E. Puthoff, "CIA Initiated Remote Viewing at Stanford Research Institute, from "The Intelligencer." *Journal of U.S. Intelligence Studies*, 12, 1, Summer 2001.

23 Reprints from *Aviation Week and Space Technology*, May 1977-Nov. 1978, McGraw Hill Inc.

24 Ibid.

25 John Pike, "The Death Beam Gap, Putting Keegan's Follies in Perspective," *The Federation of American Scientists*, Oct 1992.

26 "Soviet Fusion Research Top Secret," *Sarasota Herald-Tribune*, March 12, 1977, retrieved from https://news.google.com/newspapers?nid=1755&dat=19770312&id=c_EcAAAAIBAJ&sjid=S2cEAAAAIBAJ&pg=6736,4546187&hl=en.

27 Ibid.

28 "U.S. Classifies Lecture by Soviet Scientist," *Toledo Blade*, March 12, 1977.

29 Ibid.

30 "Kidder-Bethe Correspondence Concerning the Progressive Case," *Federation of American Scientists*, 1979.

31 Ibid.

32 Frances Fitzgerald, *Way Out There in the Blue* (New York: Simon & Schuster), . 370.

33 Personal communication.

34 http://www.nytimes.com/1983/03/27/world/andropov-says-us-spurring-race-strategic-arms-excerpts-interview-page-14.html.

35 Yonas personal diary.

36 Ibid.

37 Ibid.

38 Ibid.

39 Ibid.

40 Ibid.

41 Ibid.

42 A., Roger. "One Cheer for the Scowcroft Commission." The Heritage Foundation. April 20, 1983. Accessed May 30, 2017. http://www.heritage.org/defense/report/one-cheer-the-scowcroft-commission.

43 Yonas personal diary.

44 Ibid.

45 Ibid.

46 From Nigel Hey, "The Star Wars Enigma", Potomac Books Inc. 2006, based on discussions with the author.

47 Berkeley Breathed, Bloom County.

48 Yonas personal diary.

49 Ibid.

50 Ibid.

51 Ibid.

52 Ibid.

53 Ibid.

54 Ibid.

55 Ibid.

56 Ibid.

57 Ibid.

58 D.Nowicki and B. Muller, John McCain Report, March 01, 2007, http://archive.azcentral.com/news/election/mccain/articles/2007/03/01/20070301mccainbio-chapter6.html.

59 Dick Rose, "A Prophesy of Peace?" Unpublished, 1985, provided to the author by the Pentagon.

60 Ibid.

61 Yonas personal diary.

62 V. Mikhailov, *I am a Hawk* (Pentland Press, 1996), 84.

63 Personal communication.

64 Personal communication.

65 J. G. Mathers, "'A fly in outer space': Soviet ballistic missile defence during the Khrushchev period," January 24, 2008. http://www.tandfonline.com/doi/abs/10.1080/01402399808437716?journalCode=fjss20.

66 Tzu Sun and S. B. Griffiths, *The art of war*. Oxford University Press, 1971.

67 Yonas personal diary.

68 Gerold Yonas, "Strategic Defense Initiative: The Politics and Science of Weapons in Space," *Physics Today*, June, 1985.

69 Ashton Carter, "SDI", Office of Tech Assessment, 1986.

70 Gerold Yonas, "Can Star Wars Make Us Safe, Yes," *Science Digest*, Sept, 1985.

71 Brock Brower, "Bud-McFarlane: Semper Fi," *The New York Times Magazine*, Jan 22, 1989.

72 Jay Keyworth, private communication.

73 Personal correspondence.

74 Yonas personal diary.

75 Gus W. Weiss, *The Farewell Dossier, Duping the Soviets*, CIA Library, 2007.

76 T. C. Reed, *At the abyss: an insider's history of the Cold War* (New York: Presidio Press Book 2004).

77 Steven M. Meyer, "The USSR Use of Space", National Security Issues Symposium, 1984.

78 Ashton Carter, "The Relationship of ASAT and BMD Systems" *Weapons in Space*, 1986, 183-185.

79 P.V. Zarubin, "History of High Energy Lasers and Laser Based Systems," 2004.

80 Michael Ellman and Vladimir Kontorovich, *The Destruction of the Soviet Economic System, an Insider's History*, 58-59.

81 Evgeny P. Velikhov, ed. Shanti Blees, trans. Andrei Chakhovskoi, *Strawberries from Chernobyl* (CreateSpace Independent Publishing Platform, 2012).

82 Sidney D. Drell and George P. Shultz, "Implications of the Reykjavik Summit, 2007, 53.

83 Ibid.

84 Drell and Shultz 54

85 Drell and Shultz 58.

86 Drell and Shultz 60.

87 Oleg Baklanov Interview by Olet I. Skvortsov, Oral History of the Cold War, Feb, 1999.

88 Mikhail Gorbachev, "Turning Point at Chernobyl," *Project Syndicate*, April 14, 2006, https://www.project-syndicate.org/commentary/turning-point-at-chernobyl.

89 Roald Sagdeev, *The Making of a Soviet Scientist* (Toronto: John Wiley and Sons, 1994), 205-206.

90 Ibid.

91 Times Wire Services, "Gorbachev Tells His Grief; Czechs, Poles See 'Star Wars' Link," *Los Angeles Times*, January 29, 1986, http://articles.latimes.com/1986-01-29/news/mn-1121_1_shuttle-program.

92 George Shultz, *Turmoil and Triumph* (New York: Scribner's, 1993).

93 U.S. State Department Memorandum of Conversations, Oct 11, 12, 1986, Reykjavik.

94 Ken Adelman, *Reagan at Reykjavik, Forty Eight Hours that Ended the Cold War* (New York: Broadside Books, 2014), p.153.

95 US State Department Memorandum of Conversations, Oct 11-12, 1986, Reykjavik.

96 Ibid.

97　Oleg Baklanov, Interview with Oleg Skvorsov, Oral History of the Cold War, Feb, 1999.

98　Alexander Yakovlev, memo to Gorbachev, Feb, 1987.

99　Sidney Drell and George Shultz, *Implications of the Reykjavik Summit on its Twentieth Anniversary* (Stanford: Hoover Institution Press, 2007).

100　Ibid.

101　Klaus Gottstein, *SDI and Stability, The Role of Assumptions and Perceptions* (Baden-Baden, Germany: Nomos, 1988).

102　Yonas personal diary.

103　Ibid.

104　Ibid.

105　Ibid.

106　Ibid.

107　Klaus Gottstein, *SDI and Stability, The Role of Assumptions and Perceptions* (Baden-Baden, Germany: Nomos, 1988).

108 V.M. Zubok, *A Failed Empire* (Chapel Hill: Univ of North Carolina Press, 2007).

109 K. Lantratov, "The Star Wars which Never Happened," *Quest Magazine*, (2007), and Dwayne A. Day and Robert G. Kennedy III, "Soviet Star Wars," *Air and Space*, Jan, 2010.

110 Valery Ivanovich Boldin, Interview by Oleg I. Skvortsov, Oral History of the Cold War, Feb 1999.

111 Valery Boldin, *Ten Years That Shook the World* (Basic Books, 1940), and Boldin Interview.

112 Boldin Interview, p. 14, 30.

113 Ibid.

114 K. Lantratov, "The Star Wars which Never Happened," *Quest Magazine* (2007), and Dwayne A. Day and Robert G. Kennedy III, "Soviet Star Wars," *Air and Space*, Jan, 2010.

115 Vladimir M. Zubok, *A Failed Empire* (Chapel Hill: University of North Carolina Press, 2007), 314.

116 Francis Fitzgerald, *Way Out There in the Blue* (New York: Simon & Schuster), 370.

117 Zubok, 314.

118 Pavel Podvig, "Did Star Wars Help End the Cold War," *Russian Nuclear Forces Project*, March (2013).

119 Mikhail Gorbachev, "Gorbachev: On my Country and the World" (Cambridge University Press, 2000).

120 Ibid.

121 Zubok 344.

122 V.I. Boldin, *Ten years that shook the world: The Gorbachev era as witnessed by his chief of staff* (New York: Basic Books, 1994).

123 Curtis Johnson and Gerold Yonas, "The Wicked Future of Systems Engineering," docslide.us, Dec., 2006; and Horst Rittel and Welvin Weber, "Dilemmas in General Theory of Planning," Policy Sciences, 4, 1979, 155-169.

124 J. Gharajedaghi, "Systems Thinking, Managing Chaos and Complexity" (Elsevier, 2006).

125 US State Department Memorandum of Conversations, Oct. 11, 12, 1986, Reykjavik.

126 Frances Fitzgerald, *Way Out There in the Blue* (Simon and Schuster, 2000), 348.

127 Fitzgerald 368, 353.

128 Fitzgerald 368.

129 Alexander Yakovlev, *A Century of Violence in Soviet Russia* (New Haven: Yale University Press, 2002), 73.

130 Yakovlev 74, 238.

131 Klaus Gottstein, "SDI and Stability" (Baden-Baden: Nomos, 1988), 97.

132 Chris Johnston, "Mikhail Gorbachev: world on brink of new cold war over Ukraine," The Guardian, November 08, 2014, https://www.theguardian.com/world/2014/nov/08/gorbachev-new-cold-war-ukraine-soviet-union-us-russia.

133 Freeman Dyson, *Weapons and Hope* (New York: Harper & Row, 1984).

www.ingramcontent.com/pod-product-compliance
Lightning Source LLC
LaVergne TN
LVHW051516070426
835507LV00023B/3139